GET A GRIP ON YOUR MATH

William J. Adams

with Illustrations by

Ramunė B. Adams

KENDALL/HUNT PUBLISHING COMPANY
4050 Westmark Drive   Dubuque, Iowa 52002

To

Harold E. Lurier,

dear friend, scholar,

and teacher.

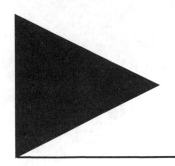

CONTENTS

Introduction   xi

PART ONE   GET A GRIP ON YOUR NUMBERS   1

1   Zero, Millions, Billions, and More   3

    Money Is Time   3
    The Curious Number Zero   7

2   Caution! Slippery Figures Sold Here   11

    Number Slinging   11
    Slippery Figures or Mathematical Proof?   12
    Soviet Defense Outlays?   14
    Assumptions Driven by Ideology: Reality's Revenge   19
    Which Came First: The Numbers or the Decision?   21
    Are American Students Really That Bad in Math and Science?   23
    Are Statistically Safe Schools Necessarily Safe?   24
    Are These Figures Comparable?   25
    Does a 999/1000 Surgical Batting Average Imply a Good
        Heart Surgeon?   26
    Political—Economic Spin or Just a Fast One?   27
    Spin Versus Counterspin   28
    More Is Less   30
    Does This Scenario Make Sense?   30
    Does Smoking a Cigarette Take Seven Minutes Off Your
        Life?   32

3   Talosian Images   35

Slippery Versus Fraudulent   35
Two for the Price of One   36
No Lofty Pinnacle   38
These Data May Give You Nightmares   38
Women's Nightmares   41
Smoking Smokescreens   43
Push Comes to Shove   45
Tailored Data   46
Doctored Data   47
Can Faking Jail Data Make You a Jail Statistic?   48
Are These Test Scores for Real?   49
Top of the Line Deception: But Who Was Deceived?   50
Little White Lies or Whoppers?   52
Number Equality By Decree?   54
Numbers That Do Not Behave "Properly"   56
Damage Control   60

4   Caution, Speed, Reliability: We Want It All   65

We Need the Data, NOW   65
The Drug That Almost Made It   66
This Drug Made It, but Questions Remain   67
Running the Stop Signs   68
Can Less Be More?   71

5   What Are the Numbers and What Do They Tell Us?   73

"Good" Numbers Would Help   73
Why the Job Loss?   74
Number Magic   75
Does This Scenario Add Up?   76
There's More Here Than Meets the Eye   77
Silence May Be Golden After All   79
Let's Get the Data   80
Let's Get the Data 2   81
Let's Get the Data 3   82

6   Statistics May Not Lie, But What Do They Say?   83

    Which Interpretation Is "Correct"?   83
    There's Less to Baseball Statistics Than Meets the Bat   84
    How Successful Was the Patriot?   85
    Political Figuring   86
    Test Scores Rise: Education Miracle or Mirage?   87
    The Reagan Economic Boom: Blessing or Disaster?   88
    More Spin and Counterspin   91
    If You Don't Agree, Shut Up!   92
    Does Baldness Cause Heart Attacks?   93
    Weighing the Data: How Credible Is Credible?   95

7   Are We On the Right Data Trail?   101

    Reliability Versus Relevance   101
    A Visit to Huxley College   102
    Which Data "Best" Reflect Airline Reliability?   103
    The National Debt: Public Nuisance or Menace?   104
    Is the Recession Over if the Statistics Say So?   106
    TV Ratings: Can You Top This?   107
    What Is the Unemployment Rate?   110
    What Is the Budget Deficit?   111

8   Coming Up with the Numbers   113

    More Easily Said Than Done   113
    Can You Trust Polls?   114
    Watch the Questions, Among Other Things   121
    Were the Numbers "Fairly" Drawn?   128
    Sexuality by the Numbers, or Not?   132
    For "Real" Accuracy, Count Them All. Right?   134
    Index Numbers in the News: How Reliable Are They?   136

9   Antidotes for Slippery Figures and Bogus Number Mongers   145

    Numbers, Numbers, Everywhere   145
    How Believable Are These Numbers? Test Questions   146
    Strengthening the Shield Against Bogus Numbers   148

PART TWO   GET A GRIP ON MATH MODELS   159

10 Mathematics for a Vacation Trip   161

    Ann's Model   161
    Andy's Model   163
    Rasa's Trip: Who's Right?   164

11 Mathematics for Business and Economics   167

    The Birth and Rebirth of Linear Programming   167
    A Tale of Two Linear Programs   168
    Is Mathematics Precise?   172
    Falling Short: The Advertising Media Selection Problem   173

12 Mathematics and the Computer   177

    How Could It Be Wrong? I Used a Computer   177
    The Computer's Right of Way   178
    If Only Computers Could Think   179

13 The Mathematics of Free Trade   181

    Some Say Yes and Some Say No   181
    What Do the Economic Models Say?   182
    The Devil's in the Assumptions   183
    What's Not in the Economic Models?   183
    What Do the Numbers Tell Us?   186
    Where Does This Leave Us?   187
    Reality's Verdict   189

14 Mathematics for Astronomy   193

    Newton Sets the Stage   193
    A New Planet Has Been Discovered, But Where Is It?   194
    An On-Target Prediction   195
    Muddle Over Method   196
    Déjà Vu   198
    Troublemakers: Mercury and Light   200

15 Mathematics and Chance   203

    The Adventures of Hasty Harry   203
    Reality Strikes Back   205
    A New Model for Harry's Die   206

16 The Mathematics of Dating   209

    Radiocarbon Dating   209
    A Tale of Fine Art Forgery   211

17 The Mathematics of Space   213

    Euclidean Geometry   213
    A Non-Euclidean Challenger   216
    Which One Is Right for Space?   221

18 A Change in the Conception of the Universe   223

    The Impact of the Non-Euclidean Challenger   223
    A Profile in Courage   225

19 The Mathematical Modeling Process as a Tool for Inquiry   229

    Overview   229
    The Final Theory?   233

20 General Lessons and Observations   235

    Fit, Fit and Fit   235
    The Consumer's Focus   236

NOTES, REFERENCES, AND READINGS OF INTEREST   239

INDEX   251

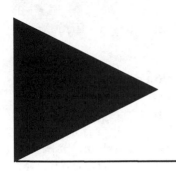

INTRODUCTION

Many of us think of mathematics as mysterious and incomprehensible, but also as precise and important. "You can't argue with mathematics; whatever career you're going to pursue, you've got to know your math," we often hear. These "wisdoms" have become generally accepted in our culture because of the views that mathematics is the epitome of precision and that it is employed in activities that affect our lives in many ways.

But this raises a number of questions: What can mathematics do for us? What are its limitations? In what sense is mathematics precise? Can mathematics be wrong even if the computations are correct? Is an argument "supported" by numbers necessarily more precise than one that is not, or might the sense of precision they convey be illusionary? What should we look for to help distinguish numbers that cast light from their bogus cousins? Our objective in this book is to demystify mathematics by addressing such questions and exploring what mathematics can and cannot do for us.

This raises another question: What technical background is needed to follow the presentation? None; curiosity and interest are sufficient. We are pleased to report that technical manipulations of the kind that strike terror in the hearts of many do not appear. We can learn a good deal about what mathematics can do for us without doing highly technical mathematics itself.

The first part of the book, Get a Grip on Your Numbers, looks at the continuous barrage of numbers we are subjected to in modern life. Numbers of a magnitude that put them beyond our comprehension— $100 billion here, $1.2 trillion there—are bandied about these days with the same abandon as slapping tennis balls to-and-fro. At the other end of the pole is the much misunderstood and mistreated number zero. The first task we set for ourselves is to get a firmer grip on such numbers.

Numbers are powerful tools of argumentation because of the sense of precision and neutrality they convey. To be told, for example, that 33% of all adult Americans are functionally illiterate is much more authoritative sounding and attention grabbing than the assertion that many adult Americans are functionally illiterate, which might be the best that could be said. The figure 33% also suggests a sense of neutrality, a haven from the vast array of self-interest and institutional biases that color the world scene.

Considering the power of numbers, it is not surprising that number slinging has become a commonly employed practice to mold opinion, disarm opposition, and convince targeted audiences to accept positions or courses of action. Properly employed, numbers help to illuminate the scene, but often there is less to them than meets the eye. All of us are potential victims of number pollution which may arise through unrealistic assumptions, number manipulation, unintentional error, or ambiguity. The best defense is provided by an awareness of how slippery numbers may arise and what we can do to spot their appearance and limit their circulation. Our objective in this part of the book is to help us develop such an awareness and sensitivity. This is undertaken by consideration of a variety of real-life situations.

The second part of the book, Get a Grip on Math Models, considers the nature of mathematical models that are set up to describe real-world situations, what we can expect to obtain from them, and their limitations. Examples considered include mathematical models for a vacation trip, for free trade, and for space.

The last section cites references to quotes and readings that might be of interest. It also provides additional commentary, usually of a more technical nature, on material discussed in the book.

As to the order in which parts and chapters are taken up, a linear order running straight through from the beginning is fine, but not necessary. Feel free to skip around as your interests, curiosity, and the chapter headings entice you.

Get a Grip on Your Math is accompanied by *Get a Firmer Grip on Your Math* which is intended for readers who wish to dig deeper. It provides questions for thought and a more in-depth discussion of some of the ideas developed in this book.

W.J.A.
R.B.A.

PART ONE

GET A GRIP ON
YOUR NUMBERS

ZERO, MILLIONS, BILLIONS, AND MORE

MONEY IS TIME

News reports inform us that David Dinkins, when mayor of New York, had given up hope of obtaining approval for $160 million in new taxes and that the City would be forced to reduce city services to close a half-billion dollar projected budget gap for 1994. We also read that Ross Perot believes that $100 billion could be saved by more efficient collection of taxes, that Federal spending for health, medicare, and agriculture for 1994 will be around $283 billion, that during the Reagan years Washington added approximately $1 trillion to the national debt, and that President Clinton outlined a spending package of $1.52 trillion for the 1994 fiscal year.

Dollar amounts running into millions, billions, and trillions are frequently bandied about these days, amounts of such overwhelming magnitude as to place them beyond imagination. Time is money, the adage goes, and by reversing direction and translating money into time it might be possible to get a better grip on the magnitude of millions, billions, and trillions of dollars. If we think of one dollar as equivalent to one second, we obtain the correspondence between time and money shown in Table 1.1.

Time Period	Seconds	Dollars
1 minute	60	60
1 hour	3600	3600
1 day	86,400	86,400
1 year	31,536,000	31.536 million

TABLE 1.1

From these time-dollar equivalents we obtain the following interesting results[1]:

$$\$1 \text{ million} = 11.57 \text{ days}$$
$$\$1 \text{ billion} = 31.71 \text{ years}$$
$$\$1 \text{ trillion} = 31,700 \text{ years}$$

The time equivalent in going from $1 million to $1 billion to $1 trillion is to go from a week and a half to just short of 32 years to just short of 32,000 years, which some may find startling (I do).

Multiplying $1 million = 11.57 days by 160 tells us that the new taxes denied Mayor Dinkins are equivalent to 1851 days, which is a bit over 5 years.

Multiplying $1 billion = 31.71 years by ½ tells us that the half-billion dollar projected budget gap faced by Mayor Dinkins is equivalent to 16 years.

It is interesting to note where larger amounts of money take us in terms of time. Multiplying $1 billion = 31.71 years, or any of the other stated equivalents, by a value of our choosing gives us money in terms of time. For the value 25 we have $25 billion, which is equivalent to 793 years. This brings us to c. 1200 A.D., a time that signals the close of a century marked by the great revival of learning and the birth of universities. *Magna Carta*, the most famous single document in the English language, seen as the origin of English constitutional freedoms, was approved in 1215. The Fourth Crusade took Constantinople in 1204, destroying all hope for the future reunification of the two halves of the Christian World split by the schism of 1054. Temujin united the Mongols and took the title Genghis Khan in 1206.

$50 billion, equivalent to 1586 years, brings us to c. 400 A.D., a time during which the Roman Empire in the West was in a state of collapse. The Visigoths sacked Rome in 410. At this time St. Augustine wrote the *City of God*, which was to profoundly influence medieval political thought. The Nicene Creed, the basic tenet of faith of the Trinitarian Christians drawn up in answer to the Arian heresy, was approved at the Second Ecumenical Council in 381. The Indian poet Kalidasa wrote the play *Shakuntala* and other Sanskrit works.

$100 billion, equivalent to 3171 years, brings us to the period centered around 1200 B.C., which saw the last stages of the Mycenean culture (c. 1600–1150 B.C.); the fall of Troy (c. 1200 B.C.), immortalized in Homer's *Iliad*; and the Hebrew Exodus from Egypt under Moses (c. 1300 B.C.).

$145 billion, equivalent to 4598 years, brings us to the period centered around 2600 B.C., which saw the Age of the Pyramids in Egypt. The Great Pyramid at Giza, begun around 2600 B.C. to house the remains of Cheops (Khufu), has an almost perfect square base with sides of 755 feet and an apex nearly 500 feet above the desert floor. Its construction consumed some 2,300,000 stone blocks weighing an average of two and a half tons each, many weighing as much as 15 tons each. In Mesopotamia the celebrated King Gilgamesh, the hero of the *Epic of Gilgamesh*[2], ruled the Sumerian city-state Uruk around 2700 B.C.

$283 billion, equivalent to almost 9000 years, brings us to Cayonu in Southern Turkey, the site at which the oldest known woven cloth, a 9000 year old piece of white linen, was found in 1988.

$1 trillion, equivalent to 31,700 years, brings us to c. 30,000 B.C., which puts us in the first Upper Paleolithic culture in Europe called the Chatelperronian period (32,000–28,500 B.C.). Chatelperronians lived in open-air settlements and caves throughout western Europe. They were not cave artists, but they engraved pebbles with pictures of animals and produced pendants of carved bone.

President Clinton's projected budget of $1.52 trillion for the 1994 fiscal year, equivalent to 48,184 years, brings us to c. 46,000 B.C., a period of transition during which Neanderthal man ceased to flourish as a type and Cro-Magnon man began to emerge.

The next level beyond the trillion which we may soon need, considering our penchant for spending money, is the quadrillion. $1 quadrillion = $1000 trillion, equivalent to 31,700,000 years, puts us in the Oligocene period of 37–25 million years ago marked by the emergence of diverse primate forms. Primates began to emerge about 65 million years ago (equivalent to $2.05 quadrillion) during the geological period called the late Cretaceous.

A number of years ago Everett McKinley Dirksen, Senator from Illinois, observed that a billion here and a billion there, and pretty soon you're talking about real money; and time too, we might add.

THE CURIOUS NUMBER ZERO

The number zero, denoted by 0, is sometimes mistakenly equated to nothing. Zero is not nothing. It is a number with the special property that zero added to any number gives us back that number; c + 0 = c, where c is any number. We may add, subtract, and multiply by zero, but we cannot divide by zero.

Division is a derivative of multiplication; the quotient c/d is defined as c multiplied by the reciprocal of d, where the reciprocal of d is that unique value whose product with d is 1. The reciprocal of 2, for example, is $\frac{1}{2}$ since their product is 1, and 4/2 means 4 times $\frac{1}{2}$, which of course is 2. But 0 has no reciprocal because there is no number whose product with 0 is 1; the product of every number with 0 is 0. Since zero has no reciprocal, c/0, which would be defined by c times the reciprocal of 0 if there were such, is not defined.

But wait, you might say. Although this analysis shows that division by zero cannot be defined in terms of the reciprocal of 0, why can't we go ahead and define division by zero in some other way? We cannot do so because no definition of division by zero compatible with the structure of our number system as a whole is possible. Any proposed definition for division by zero must lead to a contradiction within the number system, the one thing that mathematicians abhor most. Mathematicians cannot live with a system containing contradictions, which may explain why they do not become politicians.

To see why any proposed definition of division by zero must lead to a contradiction, suppose that division by zero has been defined in some way and consider the following mathematical sleight of hand which, alas, requires recall of a bit of elementary algebra.

Let x and y denote nonzero numbers such that

$$x = y.$$

Multiplying both sides of this equation by y yields

$$xy = y^2.$$

Subtracting x^2 from both sides yields

$$xy - x^2 = y^2 - x^2.$$

By factoring we obtain

$$xy - x^2 = x(y - x), \quad y^2 - x^2 = (y - x)(y + x).$$

Thus we have

$$x(y - x) = (y - x)(y + x).$$

Dividing both sides by y − x yields

$$x = y + x.$$

Since x = y, by substituting x for y in the above we obtain

$$x = x + x.$$
$$x = 2x.$$

Dividing both sides by x yields

$$1 = 2.$$

The source of this inconsistency is the step in which we divided by $y - x = 0$, under the assumption that division by zero could be defined in some way.

If $1 = 2$, then we may conclude that $1 = 2$. If a person can be found who accepts this, then he, logically speaking, should be prepared to give you two dollars for every one dollar that you give him. Whether he would be prepared to do so emotionally speaking is another matter.

Division by zero does not usually announce itself with trumpets blaring, and when division by an algebraic expression arises it must be ascertained whether division by zero could arise under certain circumstances. A spectacular case in point arose in connection with Albert Einstein's theory of relativity. At one point in his work Einstein found that certain of his equations yielded peculiar results. As a consequence he turned in another direction, a direction that subsequently had to be abandoned. These peculiarities were later explained in 1922 when the Russian physicist Alexander Friedman found that in his derivation Einstein had divided an equation by a quantity that, Friedman showed, could be zero under certain physical circumstances. This discovery and the refinement in Einstein's analysis that it brought about laid the foundation for the development of a mathematical model of the expanding universe.

While noting that division by zero is not defined, we should observe that there is nothing wrong with dividing into zero as long as we are dividing by a nonzero number. 0/2, for example, is 0 times ½, which of course is 0.

CAUTION! SLIPPERY FIGURES SOLD HERE

NUMBER SLINGING

The collection of data and determination of projections and inferences from data provide us with a powerful methodology for almost every enterprise imaginable, but this methodology is fraught with potential for misapplication, misunderstanding, and outright deception. Numerical values, because of the precision they suggest, endow statements and conclusions drawn from them with an aura of seemingly unimpeachable respectability. Consequently, the practice of number slinging has become commonplace.

The Achilles heel in this is that if the assumptions underlying the data are faulty, ideologically driven, or have been tampered with in any way, their cloud of quantitative respectability vanishes and their realism should seriously be called into question. A sound general attitude toward cited figures is one of questioning skepticism, especially when they, like the magician's rabbit, are pulled out of a hat to bolster a point of view.

SLIPPERY FIGURES OR MATHEMATICAL PROOF?

Recently the Dean of Administrative Affairs at Huxley College expressed to a friend his concern about rumors that his work is counter-productive.

Dean: "Such talk is totally unjustified and unfair."

Friend: "You're quite right. Have you pointed out that since you do no work at all, even in a leap year, your non-efforts could at worst be considered neutral. In fact, by not participating in Administration Council sessions you have made a strong positive contribution to the effectiveness of that organization."

Dean: "This sounds promising. How could I present a convincing argument?"

Friend: "There are 366 days in a leap year. Now you sleep 8 hours a day. Thus, one-third of the day is spent sleeping. You sleep one-third of the year, or 122 days. This leaves 244 days. Four hours a day are spent on rest, recreation and personal business. Thus one-sixth of the day is spent in this manner. This takes up one-sixth of the year, or 61 days, which leaves you 183 days. Of these 183 days you don't work on Saturdays or Sundays. Since there are 104 Saturdays and Sundays, this leaves you 79 days. How long do you take for meals every day?"

Dean: "It varies, but about 2 hours a day."

Friend: "Two hours a day. In a week you spend more than one-half a day eating. About 28 days are spent sitting around and eating. Twenty-eight days from 79 leaves 51 days. As a top administrator how much vacation do you get?"

Dean: "Seven weeks."

Friend: "That's 49 days from 51. This leaves 2 days; New Year's and Labor Day, which you get off. You don't even work one day in the year and your non-existent work has in fact left Huxley a much better place."

Dean: "Thanks. You always come through for me."

This variation on one of Fred Allen's routines suggests that numbers can be employed in almost any situation, sometimes with hilarious results. Alas, many of the slippery number routines we come into contact with in modern life sow confusion rather than amusement. The best form of immunity against number magic is conferred by examination of a number of slippery number routines.

SOVIET DEFENSE OUTLAYS?

The disintegration of the Soviet Union has left the United States facing some critical choices. Answers to questions about the extent to which the United States should scale back military spending are strongly influenced by analyses of the Central Intelligence Agency which indicate that the former Soviet Union had been substantially outspending the United States on military programs for the last 15 to 20 years. America's massive military buildup, especially that carried out during the Reagan years, was predicated on these assessments.

A leading American analyst of Soviet military spending, Franklyn D. Holzman of Tufts University, has consistently taken issue with the assumptions and procedures which underlie the C.I.A.'s analysis and conclusions. Writing in 1979, Holzman[1] notes:

There are many sources of possible exaggeration in the C.I.A. estimates of Soviet military expenditure relative to America's. Three of them follow:

1. Comparisons of military outlays can be made either in dollars or rubles. The C.I.A.'s published comparisons are always in dollars. Prices expressed in dollars exaggerate Soviet expenditures. This is because the Soviet armed forces have twice the personnel of America's but add only a little more new equipment each year, and because, in the words of the Director of Central Intelligence, Adm. Stansfield Turner: "In the United States manpower is relatively more expensive than hardware [while] in the Soviet Union military hardware is much more expensive than manpower." So, when the cost of the personnel of the Soviet armed forces, with their 4.5 million people—the precise number is hard to ascertain—is valued at American armed forces wages, a high Soviet defense figure, in dollar terms, results. This figure would be about $10 billion smaller if military pay were adjusted for the lower educational and training levels of Soviet soldiers. A 20 percent pay discount is regularly made by the C.I.A. in dollar comparisons involving other sectors.

2. A ruble comparison, which the C.I.A says is as valid as the dollar comparison, exaggerates American expenditures. This

is because our armed forces have more equipment per person than the Soviet forces and because equipment is relatively high priced in the Soviet Union. The C.I.A. admits this and in response to congressional questioning presented an unofficial comparison in rubles that put Soviet 1977 defense expenditures at 25 percent more than America's. This is less of a difference than the official dollar comparison, which has the Russians outspending us by 40 percent. While these two not-very-different figures satisfy Congressional interrogators, it did not satisfy economists used to such United States Soviet comparisons. Experience has shown that ruble-dollar differentials typically exceed 50 percent. Clearly, then, if the Soviet Union outspends the United States in dollars by 40 percent, one would expect the United States to equal or outspend the Russians in rubles. These C.I.A. figures, therefore, are highly suspect.

3. According to the C.I.A., the major reason why a careful ruble estimate is not made and published is that while all military equipment the Russians produce is within our technology and can be given a real dollar price, a large part of United States equipment is beyond Soviet technology and cannot be given an actual ruble price. The C.I.A. procedure in valuing American high-technology equipment is to use ruble prices "applicable to the closest substitute goods which can be produced in both economies." What this means is that the C.I.A.'s ruble calculation values this American equipment at ordinarily high Soviet ruble prices but not at what the former Director of Central Intelligence William E. Colby called prices that are so high as to be "almost uncountable." No wonder American defense expenditures priced in rubles are estimated at less than the Russians' defense expenditure. If a properly high ruble price tag could be put on our high technology, the American defense package would certainly cost the Russians more to produce than their own. It might well be that they cannot produce our defense package at any cost.

The major fallacy in the C.I.A. procedure is that the very dimension of the arms race in which America has the greatest advantage—advanced technology—and which makes most of the difference between military superiority and inferiority, is enormously undervalued.

In an analysis published in 1989 Holzman[2] observes:

> . . . over the last 15 years, the C.I.A.'s estimates have been riddled with errors and misrepresentations, all making Soviet defense expenditures appear larger than they actually are.

From 1975 through 1983, the C.I.A. said Soviet military spending was increasing 4 percent to 5 percent a year. In 1984, the agency acknowledged that it had been wrong—that the increase since 1975 had been only 2 percent a year. After conceding its error, however, the C.I.A. failed to make the appropriate adjustments in its Soviet military spending figures.

In 1982, the Soviets instituted a wide-ranging price reform, the first in 15 years. In 1986, the C.I.A. switched its estimates of Soviet defense spending, which had been in 1976 ruble prices, into the new 1982 ruble prices. Soviet defense spending took a large leap upward because, according to the C.I.A., prices of weapons had increased much faster than prices of civilian goods between 1970 and 1982. But this was all guesswork.

In November, 1987, the agency conceded that it had been unable to obtain 1982 weapons prices and, further, that there was reason to believe that weapons prices had risen more slowly than prices of civilian goods. Yet the new and much higher figures for Soviet defense expenditures were never rescinded.

In 1986, the C.I.A. made another change in established practice. Without explanation or warning, it began to include in its single estimate of Soviet military spending a number of civilian activities that are not included in estimates of United States military spending (such as civilian defense, internal security forces and civilian space activities). This made Soviet defense appear still larger relative to United States defense than it should have been.

Holzman concluded that when the C.I.A.'s errors and assumptions are corrected Soviet military spending as a share of gross national product falls almost 50 percent, from 16 percent claimed by the C.I.A. to 9 percent.

During the confirmation hearings held in the fall of 1991 on the fitness of Robert Gates to be C.I.A. Director, senior analyst Marvin Goodman and two other C.I.A. analysts testified that while serving as Deputy Director of Intelligence Gates tailored intelligence estimates to suit Administration policy in several important areas. One of the areas cited was concerned with Soviet capabilities and intentions.

In connection with this state of affairs Holzman[3] comments:

Professor Goodman's revelations come as no surprise to me; indeed, they suggest an answer to puzzles raised over the last 15 years in my studies of C.I.A. estimates of Soviet military spending.

The C.I.A. estimates have been, in my opinion, continually slanted upward to make Soviet military spending appear larger than careful analysis of the data suggests, thereby supporting the Reagan-Bush-Pentagon military buildup policies.

It was clear to me that the decisions to slant the conclusions were made at the highest levels because the techniques used

were so unacceptable on scholarly grounds. In my work on military spending, I have come to know many of the C.I.A. analysts who prepare these estimates and I believe that, had they been free of political pressure, the estimates would not have been distorted.

As to the implications of these developments for the post cold war military budget, Holzman[4] argues:

> Had there not been a C.I.A. error followed by a cover-up, our military spending might have continued to increase at a slower rate. It is estimated that for the decade 1979–1988 the United States would have spent on defense approximately $800 billion less in present-day prices. . . . The Pentagon used exaggerated estimates of Soviet military spending to help get enormous budget increases. Now that the cold war is over, this overspending should be taken into account in evaluating the Pentagon's claims to severely stretched resources.

ASSUMPTIONS DRIVEN BY IDEOLOGY: REALITY'S REVENGE

As President Jimmy Carter was preparing to leave office in January 1981, the first priority of the new Reagan Administration was to conduct a thorough overhaul of the Carter budget for fiscal 1982, which was to begin on October 1, 1981. David Stockman was appointed Director of the Office of Management and Budget and Reagan's budget team set to work in January 1981.

Stockman[5] notes: "There were three doctrines represented on the forecasting team: the monetarists, the supply siders, and the eclectics. . . . the new chairman of the Council of Economic Advisors, Murray Weidenbaum, tended toward the third approach."

In conversation with the journalist Laurence Barrett, Weidenbaum[6] commented that: "It was a forced marriage. The supply-side people insisted on [forecasting] rapid growth in real terms and the monetarists insisted on rapid progress in bringing down inflation. Each of them would go along with a set of numbers as long as their

own concern was satisfied. The monetarists weren't that concerned about growth and the supply-siders weren't that concerned about inflation."

Weidenbaum was a latecomer to the initial negotiations on economic assumptions. Barrett[7] reports that "he was so shocked at what he found that he seriously considered resigning even before unpacking the cartons in his new office. He had the same urge a few months later, when the assumptions were reviewed and retained for political reasons, though by then everyone knew they were specious." Weidenbaum stayed until the summer of 1982 because he felt that, bad as things were, the economic assumptions would have been more irresponsible than they were had he not been there.

Stockman[8] notes: "The table that follows [2.1] tells the whole story, proving that our Rosetta stone was a fake. . . . The February 1981 economic forecast eventually became known as 'Rosy Scenario.' Weidenbaum wrote the final specific numbers. But its underlying architecture—the push-pull hypothesis—was ultimately the work of a small band of ideologues."

	Real GNP Growth (%)		
Quarter	Supply-Side/ Monetarist Consensus	Final Weidenbaum Forecast	Actual Outcome
1981:4	4.0%	4.0%	−5.3%
1982:1	9.4%	5.2%	−5.5%
1982:2	7.8%	5.2%	0.9%
1982:3	6.8%	5.2%	−1.0%
1982:4	5.4%	5.2%	−1.3%

Table 2.1

WHICH CAME FIRST: THE NUMBERS OR THE DECISION?

By the early '90s there was general agreement that some military installations should be closed as a cost cutting measure. The question is, which ones? Maybe yours, but not mine, was the answer offered by local politicians as the struggle to save local bases heated up.

The Rome Laboratory in Rome, New York, a high-tech Air Force research installation, came under scrutiny. An analysis carried out in October of 1994 of a proposal to close the Lab and move its facilities to Hanscom Air Force base in Massachusetts led to the conclusion that it would cost $133.8 million to move the Lab with annual savings of $1.5 million being realized. It would take more than 100 years to recover the cost of the move at this rate. The Pentagon did not recommend closing the Rome Lab.

In February of 1995 the Air Force released an analysis of a proposal to close the Rome Lab and move 60 percent of its operations to Hanscom and 10 percent to the Army base in Fort Monmouth, New Jersey. It led to the conclusion that relocating the Lab would cost $52.8 million with annual savings of $11.5 million. The cost of the

move would be recovered in four years at this rate. The Pentagon recommended closing the Rome Lab.

In May 1995 another analysis led to the conclusion that relocating the Lab as described would cost $79.2 million with annual savings of $13 million. The cost of the move would be recovered in six years at this rate. Again the Pentagon recommended closure.

Which numbers, if any, are realistic? They arise from different assumptions, which means that the question of number realism falls back on the question of assumption realism. Did politics play a role in the decision to close the Rome base? Senator Alfonse D'Amato of New York believes that the answer is yes. They reached a back room political decision to close the Rome base and generated numbers that would justify that decision, Senator D'Amato and other New York officials believe. Analysts with the Air Force deny this, and argue that the third analysis is the most comprehensive and realistic. In April of 1995 the General Accounting Office criticized the Air Force for providing insufficient documentation for the analyses that prompted decisions to close bases. This takes us back to square one, the nature of the assumptions.

AFTERMATH. In June of 1995 a Presidential commission reversed the Pentagon's closure recommendation. It gave a 13 year estimate for the time it would take to recoup the cost of relocating the Lab, which it felt was too long to justify its closure. Is 13 years more realistic than the previous estimates of 100+, 4 and 6 years? Again, this takes us back to the assumptions.

ARE AMERICAN STUDENTS REALLY THAT BAD IN MATH AND SCIENCE?

Every few years another study appears which shows again that American students are worse in math and science than their counterparts in even the poorest countries. But are they being compared to their counterparts? Some say no. Without denying that there is much to be improved in American education, a growing number of critics

have argued that the test results are flawed because American students in total are consistently being compared with the elite students of other countries.

ARE STATISTICALLY SAFE SCHOOLS NECESSARILY SAFE?

In July 1986 the New York City Board of Education issued a list of its most dangerous schools based on incident and crime reports that it had received. One respondent[9] notes that he never felt unsafe or threatened in teaching at the fifth listed "most dangerous" junior high school. Another respondent[10] comments: "I believe that the scorecard . . . names not the most dangerous schools but the schools whose administrators have the courage to report what is really happening."

ARE THESE FIGURES COMPARABLE?

According to a study published by the American Federation of Teachers in 1993 American school teachers make less money than their peers in many other countries. The top salaries for American high school teachers average $38,000 a year as opposed to $43,000 in Germany, $45,000 in Japan, $47,000 in Ottawa, and $70,000 in Geneva, we are told. Such gaps, some argue, make clear the high priority given to the teaching profession in other lands compared to our own attitude toward the teaching profession.

But are the salaries comparable? Two questions arise: Is it meaningful to use a dollar standard at market exchange rates to compare economies? In making such a comparison, how far do the compared salaries go in terms of cost of living and tax costs?

DOES A 999/1000 SURGICAL BATTING AVERAGE IMPLY A GOOD HEART SURGEON?

Heart operations have become commonplace these days and a crucial question facing a person in need of heart surgery is: How good is my surgeon? As he recalls in his book *A Bus of My Own*, this question faced Jim Lehrer, co-anchor of *The MacNeil/Lehrer Newshour*, after his heart attack in 1983. "What is your win-loss record?" Lehrer asked the surgeon. "Ninety-eight percent or better," the surgeon replied.[11]

Many of us would feel embarrassed about asking a "professional" such a question, but faced by the necessity of a heart operation (or operation of any kind for that matter) it's better to be embarrassed and alive than dead. Jim Lehrer was clearly on the right track, but how revealing are surgical batting averages? Unlike ball players who must face all pitchers who come their way, surgeons can choose those whom they are willing to operate on. Those who make it a practice to avoid the tough cases will emerge with a high surgical batting average. Clearly there is more to a "good" surgeon than a success ratio by itself suggests.

POLITICAL—ECONOMIC SPIN OR JUST A FAST ONE?

In the winter of 1993, the Clinton administration was fighting hard for a big new jobs program. In early March, the February payroll figures appeared showing that the economy had created 365,000 new jobs in February. This, the biggest monthly gain in years, took the sails out of the case for a big new jobs program, unless the figure could be "properly" accounted for. Secretary of Labor Robert Reich attempted to meet the challenge by calling a special news conference at which he claimed that the Administration's job stimulus program had not been undercut because 90 percent of the new jobs created were of a part-time nature. Alas, there is no way to tell what proportion of the payroll jobs were part-time from the payroll data since the

Labor Department does not ask businesses whether their payroll jobs are full or part-time when their surveys are taken. The 90 percent figure was obtained by going to another survey, called the household survey, which asks individuals surveyed whether they worked in the previous month and, if so, whether it was of a full or part-time nature. But the results obtained from the household survey are not applicable to those obtained from the payroll survey so that the 90 percent figure remains without a basis.

The difficulty was obvious to professional economists and statisticians in the Labor Department and they drafted an internal memorandum explaining why the Labor Secretary could not conclude what he had concluded to answer the telephone inquiries that were being received.

SPIN VERSUS COUNTERSPIN

Speaking on television on Tuesday night of 3 August 1993, President Clinton described the budget legislation then before Congress as "the largest deficit reduction in history." Almost immediately after the President spoke, Senator Robert Dole, Republican leader in the Senate, described the legislation as "the largest tax increase in world history."

Who is right? Neither; when the dollar amounts are adjusted for inflation so that dollar comparisons are meaningful, 1993's budget bill is neither the biggest reduction measure nor the biggest tax increase in recent years. In 1993 dollars, the bill would lower the annual deficit by a projected total of $496 billion over five years; $241 billion of this would come from tax increases. The bill signed by George Bush in 1990 contained $532 billion in deficit reduction in terms of 1993 dollars. The bill signed by Ronald Reagan in 1982 raised taxes by $286 billion over five years in terms of 1993 dollars.[12]

MORE IS LESS

President Clinton's initial commitment was to reduce spending on U.S. intelligence programs by $7 billion over four years. But then he decided to ask Congress for a $400 million increase for 1994. Since Congress had in 1993 authorized about $400 million more for intelligence operations than it finally appropriated, the Administration took the position that its proposed increase was not an increase but a freeze.

DOES THIS SCENARIO MAKE SENSE?

In the spring of 480 B.C. the Persian King Xerxes launched against Greece the largest military operation the world had seen to that point. But how large? The Greek historian Herodotus gives the fighting force as 2,641,610 men. After considering the service train that went with them, Herodotus put the total force at 5,283,220 men. But, he adds:

> no one could give the exact number of women who baked the bread, or of the concubines, or the eunuchs, or the transport animals and baggage-carrying cattle and Indian dogs that came with the army—of all these creatures no one could count the numbers, they were so large.[13]

The invasion force was large, no doubt, but 5,283,220 men plus others? This exceeds by far the entire population of Greece at that time.[14] The logistic problems of maintaining such a force are, to put it mildly, considerable.

The war ended disastrously for the Persians. A small Spartan force of 300 sacrificed itself at Thermopylae and gained time for Greece. The decisive battle was fought off the island of Salamis. Xerxes watched in a rage as the smaller Greek fleet trapped, out-maneuvered, and destroyed the Persian fleet.

The Persian War was a watershed event in European history. The West had been saved from Asiatic domination and allowed to develop its own institutions.

DOES SMOKING A CIGARETTE TAKE SEVEN MINUTES OFF YOUR LIFE?

Every cigarette steals seven minutes of a smoker's life, stated the Centers for Disease Control and Prevention and reported by *The New York Times*.[15] This figure was obtained by focusing on 418,690 deaths in the United States in 1990 that were viewed as premature and directly attributed to cigarette smoking. These deaths added up to 5.04 million years of projected life (presumably obtained in terms of life expectancy values). Dividing this projected number of years lost by the estimated average number of cigarettes these people had smoked before dying and estimating the time it takes, on average, to finish a cigarette as seven minutes yields a value of seven minutes. The interpretation offered is that every cigarette steals seven minutes of a smoker's life.

It is certainly a graphic view of the consequences of smoking, but is it realistic? The question is: Does it work that way? The evidence is clear that cigarette smoking adversely affects the health of smokers, but in different ways and with different degrees of severity when smokers are considered individually. Many of us know cigarette smokers who have been puffing away since their teens and are still going strong well into their 80s and 90s. They have lived beyond expected life spans and some might even argue (absurdly, needless to say) that smoking a cigarette added a certain number of minutes to a smoker's life for smokers in this group.

ZENO'S PARADOX. The Greek philosopher Zeno of Elea (c. 450 B.C.) is mainly remembered for his paradoxes of motion, one of which is that a hare (or Achilles, if you prefer) will never overtake the tortoise which he pursues because the hare must first reach the point from which the tortoise started, so that the slower tortoise must always be some distance ahead of the faster hare. This is most impressive when cast in mathematical form, which can be done in the following way. The hare can run 16 feet in 1 second and the tortoise can run 4 feet in 1 second, let us say. The hare was to race the tortoise, but the tortoise was to be given a handicap of 16 feet. Now, by the time the hare had run the 16 feet from his starting point to where the tortoise started, the tortoise would have run ahead 4 feet; by the time the hare would have moved ahead the 4 feet, the tortoise would no longer be there, having moved ahead 1 foot, and so on. Thus, the hare can never catch the tortoise.

Although the reasoning, especially when cast in terms of numbers, may sound very convincing, we all know that hares catch and surpass tortoises, so that obviously something is wrong with the analysis. Many situations are of this kind; the conclusion doesn't seem realistic, but the mathematics seems convincing at first sight and you cannot pinpoint a flaw. When you encounter such a situation think of Zeno's paradox, which helps, I find, to put it into perspective.

The smoker's loss of seven minutes of life per cigarette smoked may initially seem to be this sort of situation, but let us note that the starting point of the analysis is concerned with cigarette smokers who die prematurely as measured by life expectancy values, and the conclusion concerns all cigarette smokers, whether they die prematurely in terms of these measures or not.

3 TALOSIAN IMAGES

SLIPPERY VERSUS FRAUDULENT

In the Star Trek adventure *The Menagerie* Captain James T. Kirk and the Enterprise encounter the Talosians of Talos IV. The Talosians have the power to create enticing illusions which, to some, are preferable to reality. To achieve Talosian images on Earth, image makers may employ Talosian figures consisting of data that have been altered, selected, or obtained under circumstances that have been compromised. In a frank conversation between television inter-

views that was inadvertently broadcasted across his country, Brazil's finance minister Rubens Ricupero expressed the sentiments of many kindred spirits when he confessed of economic indicators: "I have no scruples, what is good we take advantage of. What is bad, we hide"[1]

The difference between Talosian figures and slippery ones, which may be termed semi-Talosian, is one of degree. Arguably, a case considered under one heading might warrant placement under the other. As the following cases make clear, Talosian figures, like their semi-Talosian cousins, may arise from the widest imaginable spectrum of activities. It is not the activity but the human condition that makes the decisive difference.

TWO FOR THE PRICE OF ONE

On February 4, 1993 a state court jury in Atlanta held General Motors responsible for the death of a Georgia teenager in an accident in which his 1985 GMC Sierra pickup truck burst into flames when it was struck from the side. The jury determined that G.M. knew the truck's design was flawed, but kept the danger from the public. It

awarded $4.2 million in actual damages and $101.2 million in puni-
tive damages to the parents of the boy who was killed. Key testimony
was provided by Ronald Elwell, a former G.M. safety engineer, who
testified that the company withheld crash test data from him until
1983 and that he continued to defend G.M. until 1989 out of fear of
losing his job.

Citing technical points, a Georgia appellate court overturned the
verdict in June 1994, but held that the evidence submitted showed
that G.M. was aware that the gas tanks posed a hazard but did nothing
to make them safer.

In October 1994 Transportation Secretary Federico Peña announced an initial finding that G.M.'s pickup trucks posed a fire
hazard when struck from the side and further stated that G.M. management at the time appears to have made a decision favoring sales
over safety. In December 1994 Mr. Peña announced a settlement with
G.M. that would end the Federal investigation in exchange for G.M.'s
promise to contribute $51.3 million to auto safety programs.

After losing the court case G.M. promptly announced its intention
to sue NBC for fudging the scene on its Dateline NBC program of
November 17, 1992. This program showed a simulated crash involving the type of truck that was at the heart of the suit in progress

against G.M. The fuel tank of one of two trucks tested exploded. Viewers were not informed, however, that ignition devices had been taped to the trucks to ignite a fire if the simulated crash did not produce the desired result. To forestall a costly suit NBC apologized on the air for its manipulation of the tests.

A double play in manipulation, and the consumer is out twice.

NO LOFTY PINNACLE

In 1986 Dr. David Baltimore, renowned recipient of the 1975 Nobel Prize for medicine, Dr. Thereza Imanishi-Kari of Tufts University, and others published a research paper about genetic influences on the immune system which was hailed as a major discovery in immunology. The research underlying the paper was carried out under Baltimore's general direction with Imanishi-Kari as laboratory director. About two months after the paper was published some of the findings on which it was based were called into question by Dr. Margot O'Toole, a post-doctoral research assistant who had worked on the project.

Both her allegations and O'Toole herself were dismissed, and Baltimore vigorously defended the paper and stonewalled other inquiries. In 1991 a National Institute of Health investigation concluded that data had indeed been fabricated. Dr. O'Toole, who paid a heavy price for her integrity and courage, was vindicated. Dr. Baltimore, who was subsequently appointed president of Rockefeller University, was forced to resign this position because of his role in the scandal.

THESE DATA MAY GIVE YOU NIGHTMARES

Halcion, manufactured by the Upjohn Company and introduced in the United States in 1983 is one of the world's best known sleeping pills. Its main advantage over competing products, Upjohn has claimed, is in encouraging nighttime sleep without daytime drowsiness.

How safe is Halcion? It received Food and Drug Administration approval and its manufacturer claims that it is just as safe as other drugs of its kind. Dissenters argue that Halcion is more likely to cause symptoms such as amnesia, paranoia, and depression and that Upjohn engaged in data manipulation to conceal its side effects. This view emerged from a law suit filed by Ilo Grundberg, who killed her mother the day before her mother's 83rd birthday and placed a birthday card in her hand. Mrs. Grundberg claimed that Halcion had made her psychotic, and charges against her were eventually dismissed. (The author William Styron describes his experience with Halcion in his book *Darkness Visible: A Memoir of Madness*.[2]) Upjohn settled the lawsuit with Mrs. Grundberg before it was to go to trial in August 1991, but in preparation for the suit it had to make available a good deal of data about Halcion to the plaintiff's attorneys.

Dr. Ian Oswald, who was head of the department of psychiatry at the University of Edinburgh and spent 30 years doing research on sleep, was obtained as an expert witness. Dr. Oswald spent two years going over Upjohn's data and concluded that Upjohn had known about the extent of the drug's adverse effects for 20 years and concealed these data. He concluded that "the whole thing had been one long fraud."[3] Dr. Graham Dukes, former medical director of the Dutch drug regulatory agency, who examined some of Upjohn's data, believed that the data on Halcion had been organized in such a way as to minimize the drug's adverse effects and that this could not have occurred accidentally.

In reaction to the criticisms voiced, Britain, the Netherlands and Belgium were led to remove the sleeping pill from the market. A report issued in April 1994 by F.D.A. investigators stated that the Upjohn Company had engaged in ongoing misconduct with Halcion. The F.D.A. will investigate, it was announced.

WOMEN'S NIGHTMARES

Did the Dow Corning Corporation, the leading manufacturer of silicone gel breast implants, "honestly" report test results concerning the safety of their product? Many would say no. In one case internal memos and reports suggest that important results of a study involving dogs were incorrectly reported. Four dogs given implants in the study were reported in a medical journal as having remained in good health, when in fact one had died, one had developed a benign tumor, and the other two had suffered from a persistent inflammation.

In 1994 lawyers representing breast implant patients found a 1975 study by Dow showing that D4, a particular type of silicone gel which is a building block for other types of silicone, seriously impairs the immune system of mice. Dow Corning has steadfastly denied any intentional wrong doing and has argued that its accusers have engaged in misrepresentation.

The Dow Corning case illustrates how "narrow" scientific perspectives based on one's discipline can lead to an incomplete overall perspective. In the mid-1960s two scientific teams were set up by Dow Corning to study the behavior of silicone compounds. One unit, composed primarily of biologists, was interested in biologically active silicone compounds. They found that low molecular weight silicones stimulate the immune system. Inert compounds had no interest for them. The second unit, composed primarily of chemists and toxicologists and located down the hall from the biologists, focused on inert forms of silicone for medical uses such as implants. It's almost like having surgeons and chiropractors next door to each other, each viewing a patient in very different terms. Ultimately, Corning's business decision makers paid more attention to their chemists than their biologists.

The eventual outcome was that more than one million American women received silicone gel implants. Thousands of women have filed lawsuits claiming that the product caused them to develop auto-immune disease. Further developments are discussed in "Weighing the Data: How Credible Is Credible?" (Ch. 6.)

SMOKING SMOKESCREENS

In the late 1970s the Philip Morris Company set up a highly secret research unit to study nicotine and its effects on the body. Scientists hired to work in the unit had to sign an agreement which bound them not to discuss or publish their research without Philip Morris's permission. Two key researchers, Drs. Victor DeNoble and Paul Mele, who worked in the unit from 1980 until it was closed in 1984, painted a startling picture of a hush, hush operation and data suppression in testimony given before a Congressional hearing on April 28, 1994.

Among the findings suppressed were data indicating that nicotine was addictive to laboratory rats; that another common product of tobacco combustion, acetaldehyde, was addictive; that a combination of nicotine and acetaldehyde sharply increases the addictive power of each; the discovery of a nicotine-like substance, called 2′ (read two-prime) methylnicotine, which causes animals to experience the same "high" as nicotine, but without toxic effects on the heart; and the finding that secondhand smoke has a toxic effect on plants.

Although the findings were preliminary, they were excited about where they might lead, the scientists stated, but then it all came to an abrupt end in a reaffirmation of the adage: If you don't want answers, don't ask questions. On April 5, 1984 they were told to halt their studies, kill the laboratory animals, and turn in their security badges by morning, the scientists testified.

A SMOKING GUN. Internal documents from the Brown & Williamson Tobacco Corporation going back to the early 1960s reveal a very different picture from the public posture that we have become accustomed to being presented. In 1963, when Dr. Luther Terry was preparing the first Surgeon General's report warning against cigarette smoking as a health hazard, an internal debate raged over what research data in the company's possession should be disclosed to the Surgeon General and the public position that should be adopted.

In July 1963 Addison Yeaman, then general counsel for the company, suggested that the company "accept its responsibility," inform the Surgeon General about what it knew about the hazards of cigarette smoking, and pursue research to develop safer cigarettes.

This counsel was not followed and the tobacco industry pursued a public relations strategy emphasizing the smoker's freedom of choice and the glamour and sophistication of smoking while denying that health hazards had been proved.

PUSH COMES TO SHOVE

Dr. Daniel Tripodi, an immunologist, achieved a high career point in 1988 when he moved from director of biotechnology at Johnson & Johnson to vice president of research and development at Therakos, a company subsidiary. Therakos had been created to develop products based on a promising medical procedure, called photopheresis, which involved removing blood from patients, exposing the white cells to ultraviolet radiation, and then returning them to the body. Therakos developed a photopheresis machine, called Photoceptor, which received Federal Drug Administration approval in the spring of 1988. Dr. Tripodi's role was to develop the next generation photopheresis machine, called Centrinet. His first year at Therakos went well, Dr. Tripodi relates, and he was generously rewarded. And then it abruptly crashed.

The issue which led to disagreement with his non-scientist boss centered on the reliability of available data on Centrinet's performance. Tripodi felt that additional data based on clinical trials were absolutely needed before seeking F.D.A. approval because Centrinet's design was so different from Photoceptor's. Higher management's position was that data from the laboratory trials conducted with Centrinet were sufficient and that additional clinical trials were unnecessary because of Centrinet's similarity to Photoceptor. Additional clinical trials are costly and time consuming and at stake was a potential $12 million market.

Dr. Tripodi was dismissed. He says that it was because he resisted pressure to submit misleading test data to the F.D.A. to obtain quick approval for Centrinet. The Company's position is that he was dismissed because he had fallen behind in the development of Centrinet and was not a good manager.

In the Federal court case that followed Dr. Tripodi's dismissal, the jury found that Therakos did not have good reason to dismiss him

and that his dismissal violated Johnson & Johnson's corporate "Credo" which, in part, states that management must be just and ethical and that employees must feel free to make suggestions and have a sense of security in their jobs. Dr. Tripodi was awarded $434,000.

TAILORED DATA

In early February 1993 shareholders of the Leslie Fay Companies took legal action against the management and directors of the women's apparel maker after it announced that accounting irregularities might eliminate its 1992 profits. "Creative accounting" was used to inflate the value of inventories while lowering the reported costs for producing merchandise. A preliminary audit revealed that 1991 earnings were $17 million as opposed to the reported $29 million. The anticipated earnings of $24 to $25 million for 1992 was in reality a $13.7 million loss, not including the cost of the investigation itself. The company's comptroller was apparently so engaged in tailoring the data that he failed to reconcile key financial statements, which brought the situation to light.

Fortunately for the rest of us, it's not easy to be a successful big time data manipulator.

DOCTORED DATA

In June 1993 Empire Blue Cross and Blue Shield admitted that the company had repeatedly filed false data about its finances to the New York State Insurance Department. During the period in question, 1989 through 1992, Empire obtained a series of large rate increases.

How could this go on for so long? According to a United States Senate subcommittee report, it was a case of incompetence all around. Empire's managers were inept, its board of directors was asleep, the State Insurance Department was nonresponsive to obvious signs of trouble, and an independent audit carried out by a prominent accounting firm was seriously flawed.

CAN FAKING JAIL DATA MAKE YOU A JAIL STATISTIC?

Many of us would hope that the answer is yes. The specific case in question came to light in April 1992 when New York City's Department of Investigation issued a report stating that New York City jail supervisors and guards falsified records in 1991 in at least 85 cases, and perhaps hundreds more, to conceal the amount of time it took to provide inmates with "proper" detention facilities defined by court order.

The report observed that records were falsified to get around a court order requiring that a new inmate be assigned a bed in a cell or dormitory within 24 hours of being placed in the custody of the Corrections Department.

ARE THESE TEST SCORES FOR REAL?

When the third graders in the Cherokee School in an affluent suburb of Chicago had substantially better test scores than the third graders in the district's other three schools, very competitive and upset parents wanted to know why. In the investigation that followed, one teacher testified that the principal told her to change answers on completed tests. Another testified that the principal instructed her to erase answers on an incomplete math test so that the test would not be counted in the student's or school's average.

After an initial suspension from her administrative duties, the principal was "demoted" to a teaching position. Ethics, honesty, and integrity are not formal subjects in most elementary schools and one may argue the pros and cons of making it so. But in this case it would be uncertain as to what message would be received in doing so. Do we behave as teacher says or as teacher does?

TOP OF THE LINE DECEPTION: BUT WHO WAS DECEIVED?

In August 1993 former Administration officials stated that a key 1984 test in the "Star Wars" missile interception project had been rigged. After three tests in the series had failed, the fourth was faked, a former Reagan Administration official stated. It was crucial that the fourth test succeed, a scientist connected with the project said; otherwise we would lose hundreds of millions of dollars in Congress. "We put a beacon with a certain frequency on the target vehicle. On the interceptor, we had a receiver. . . . The hit looked beautiful, so Congress didn't ask any questions."[4] It was nearly seven years before project researchers were able to hit a missile with a missile again, and then with no consistency. It was claimed that the intent was to deceive the Soviets about the progress of the "Star Wars" system and induce them to spend tens of billions to counter the American effort, thereby draining their economy.

Major General Eugene Fox, deputy program manager for the test, commented that the target missile was artificially heated, which made it 10 times more visible to the interceptor missile than an actual Soviet missile would have been, but denied that Congress was misled about the experiment. He also said that an explosive charge had been placed on the target missile so that if it were nipped it could be better seen.[5] John Pike, director of space policy at the Federation of American Scientists, observed that from General Fox's comments it appears that the test was rigged in at least two ways and that the debate was on how it was rigged, not whether it was rigged.

At a news conference held on 9 September 1993 Defense Secretary Les Aspin, referring to a hasty Pentagon inquiry about the 1984 test, denied accusations that Congress had been misled about the success of the test.[6] Senator David Pryor of Arkansas disputed Mr. Aspin's findings. "The Pentagon is like the student grading its own exam paper in these operational test situations," he observed.

In 1992 the General Accounting Office audited seven "Star Wars" tests conducted between 1990 and 1992. It found that four of the test results described to Congress as successes were false whereas the three tests that were described as complete or partial failures were correct.

LITTLE WHITE LIES OR WHOPPERS?

In March of 1994 it was revealed that a Federal investigation had found that Dr. Roger Poisson of St. Luc's Hospital in Montreal had falsified data in a study published in 1989 that helped change the way breast cancer is treated. The conclusion that this and other studies supported is that lumpectomy, involving removal of only the area of cancerous tissue, followed by radiation, is as effective as total mastectomy in preventing the spread of the cancer in women with early breast cancer.

The flawed data involved more than 100 of 1511 patients that Dr. Poisson had enlisted in his part of the large scale N.S.A.B.P. (National Surgical Adjuvant Breast Project) study. The data concerning the 100 odd patients were flawed in that these patients were included in the study but did not satisfy the criteria for inclusion. Dr. Poisson saw his lapses as trivial violations of the rules and insisted that he had done little more that tell "white lies" which do not alter the conclusions reached by the study.

The question is, if a study is not consistent with its own criteria, how can we have confidence in its conclusions? This is the problem that Dr. Poisson's "white lies," as he views them, raised. His study contributed about 16 percent of the data to the overall N.S.A.B.P. study. A reexamination of the data, excluding Dr. Poisson's results, led to the conclusion that the conclusion reached in the N.S.A.B.P. study still holds.

WHOPPERS, FULL-BLOWN. In August of 1993 Dr. Barry Garfinkel, a widely known child psychiatrist, was convicted by a Federal jury in Minnesota of falsifying data in a study concerning Anafranil, an anti-depressant drug that the University of Minnesota engaged him to test on patients with obsessive-compulsive disorders.

Evidence was received that Dr. Garfinkel went so far as to falsify the very existence of patient visits.

NUMBER EQUALITY BY DECREE?

This is a story which illustrates how half-baked knowledge, coupled with a well-intentioned desire to share the light, passed to those with unbaked knowledge but with authority to make law, can lead to legally sanctioned Talosian images. This situation, alas, is not unique to mathematics alone.

Our story has roots in what is called the problem of squaring the circle, which goes back to ancient times. The problem of squaring the circle is to construct a square whose area is equal to that of a given circle, where the construction may be carried out with the aid of ruler and compasses only. It is known that the problem has no solution. This fact, however, has not deterred circle-squarers, as they are called, who steadfastly believe that the problem can be solved (it's just that mathematicians haven't been clever enough to do so) and that they have succeeded in doing so. Circle-squarers exist in every land in every age.

Edwin J. Goodman, M.D., of Solitude, Posey County, Indiana was one of America's circle-squarers. Dr. Goodman's "solution" of the problem leads to the conclusion that π (the ratio of the circumference of a circle to its diameter) is equal to $16/\sqrt{3}$, which is approximately 9.24. ($\sqrt{3}$ is that number which multiplied by itself yields 3; $\sqrt{3}$ is approximately 1.732, George Washington's birth year with a decimal point.) Since π is approximately 3.14, the good doctor's estimate is considerably off the mark.

Armed with the courage of his confusion and a zealous desire to share his discovery with fellow Indianans, Dr. Goodman succeeded in having a bill on his discovery introduced in the Indiana House of Representatives on January 18, 1897. It was listed as House Bill No. 246, entitled "A Bill introducing a new Mathematical Truth." Its preamble noted:

> A bill for an act introducing a new mathematical truth and offered as a contribution to education to be used only by the state of Indiana free of cost by paying any royalties whatever on the same, provided it is accepted and adopted by the official action of the legislature in 1897.

Prophetically, the bill was initially referred to the House Committee on Swamp Lands, which passed it on to the Committee of Education which, in turn, recommended House approval. In the course of House debate a former teacher stood squarely, if not soundly, behind the bill with these words:

> The case is perfectly simple. If we pass this bill which establishes a new and correct value of π, the author offers our state without cost the use of this discovery and its publication in our school textbooks, while everyone else must pay him a royalty.

The House passed the bill by a vote of 67 to 0 on February 5, 1897.

The bill then went to the Senate, where it was referred to the Committee on Temperance. It too recommended the bill's passage

and it passed the first reading in the Senate. Dr. Goodman's well-intentioned desire to purge his fellow Indianans of the errors of their mathematical ways seemed to be within reach when these goings-on came to the attention of a member of the mathematics department of Purdue University, Professor C. A. Waldo, who was visiting the state capital on other business. Professor Waldo was able to inform the senators about the nature of what was being proposed and on its second reading, February 12, 1897, the Senate voted to postpone consideration of the bill indefinitely.

NUMBERS THAT DO NOT BEHAVE "PROPERLY"

From our education and experience we have come to take certain properties of numbers for granted. We expect to be able to add, subtract, multiply and divide them, at the very least, and numbers which do not lend themselves to these operations can only be described as ill-behaved.

To see the likes of such creatures let us turn to a questionnaire sent by the student government of Huxley College to each student at the end of the academic year.

Huxley College Student Government Questionnaire

Personal Data: Please indicate the following:

1. Sex: (1) Male, (2) Female

2. Marital status: (1) Married, (2) Single, (3) Separated, (4) Divorced

3. Age: (1) Under 18, (2) 18–19, (3) 20–24, (4) Over 24

4. Work status: (1) Employed full time, (2) Employed part-time, (3) Not employed

5. Highest SAT (Scholastic Aptitude Test) score (total): (1) Under 900, (2) 900–1 000, (3) 1001–1100, (4) Over 1100

University Related Data

6. Class designation: (1) Freshman, (2) Sophomore, (3) Junior, (4) Senior

7. Area of major: (1) Business, (2) Humanities, (3) Science, (4) Computer Science, (5) Mathematics, (6) Education, (7) Engineering, (8) Undecided

8. Distance of residence from campus: (1) Under 1 mile, (2) 1–3 miles, (3) Over 3 miles

9. Preferred class room temperature (Fahrenheit): (1) 66–68, (2) 69–70, (3) 71–72, (4) 73–74

10. Satisfaction with teaching quality this past academic year (smallest number indicates highest degree of satisfaction): 1 2 3 4 5

The data arising from questions 1, 2, 4, and 7 are said to have a **nominal scale** because the responses express categories for which no ordering is implied. In question 7, for example, eight categories are cited, but there is no ordering implied as stated which would put any one category ahead of any other. Numbers are assigned to the categories to help record the responses, but this is their only role. Nominal scaling is the weakest form of measurement.

The data arising from questions 6 and 10 are said to have an **ordinal scale** because there is an implied order which allows us to speak about one category being better or preferable to another. In question 10, a rating of 1 is better than or superior to a rating of 2 or any of the other ratings. Ordinal scaling is stronger than nominal scaling, but it is still weak. We cannot meaningfully talk about differences of ratings in an ordinal scale because there are many ways to choose such numbers and differences depend on which numbers are chosen. In question 10 the ratings could just as well have been 2, 4, 6, 8, 10 or A, B, C, D, F. Numbers (perhaps better called numerals) are employed in an ordinal scale to indicate ranking, but this is as far as we can legitimately go. It is meaningless to add, subtract, multiply and divide by ordinal numbers.

The data arising from questions 5 and 9 are said to be measured on an **interval scale**. Interval data are more than rankings; they are numerical values for which differences are meaningful. With temperature readings, for example, there is a meaningful difference of one degree between each unit. One property lacking with an interval scale is that there is no true zero. The 0° Fahrenheit mark is an artificially chosen value, as is the 0° centigrade (Celsius) mark; zero degrees Fahrenheit or Celsius does not mean no heat. Ratios have no meaning in an interval scale. It is meaningless to say, for example, that an object with a temperature of 60°F is twice as hot as one with a temperature of 30°F.

The data arising from questions 3 and 8 are said to be measured on a **ratio scale**. Ratio scale data have the properties of interval scale data and in addition there is a true zero point so that it is meaningful to consider ratios of measurements. A distance of 4 miles, for example, is twice the distance of 2 miles; an age of 30 years is three times the age of 10 years. Temperature measured in the Kelvin scale, where there is an absolute zero point, is on a ratio scale.

(Absolute zero may be defined as the temperature at which the volume of a gas seems to become zero when the data in a plot of volume versus temperature for the gas are extrapolated.) Ratio scale numbers are "properly" behaved in that we can meaningfully perform with them the standard arithmetic operations that we have been brought up with.

Failure to recognize the type of data scale being employed from the context of its usage leads to widespread numerical Talosian images. Consider, for example, the following statements:

1. Al is twice as intelligent as Bob.

2. My social security number is better than yours because it's larger.

3. Donna is twice as good at tennis as Lisa since Donna is ranked third and Lisa is ranked sixth.

4. With the temperature reaching 84°F today, it's twice as warm today as it was last week when the temperature fell to 42°F.

5. Professor Arness, with a student ranking of 4 from one class and a student ranking of 2 from another, is twice as good in one class as the other.

6. Jim, who got 90 on his last math exam, knows twice as much math as Harry, who got 45.

Each one communicates a Talosian image because each is meaningless as a quantitative statement. Statements (1) (4) and (6) are based on an interval scale for which ratios are not meaningful. Statements (3) and (5) are based on an ordinal scale for which ratios are not meaningful. Statement (2) is based on a nominal scale for which no ordering of the categories is implied.

DAMAGE CONTROL

Let us put ourselves in the position of a narrow self-interest group dedicated to warding off troublesome data leaks, containing damage from unfavorable publicity and judgments, and cloaking its clients or firm with an aura of respectability. What guidelines should we follow?

This summary of spin techniques contains nothing new for practitioners of spin; it is for the rest of us who would like a glimpse of the tools of the spin trade.

1. Obtain Pledges of Confidentiality. Require your employees and the employees of service organizations who have access to "sensitive" material to take a pledge of confidentiality regarding such.

In situations concerning trade secrets, advertising campaigns, legitimate matters of national security, and the like, this is sensible and proper procedure. It does, however, have another cutting edge that was revealed in the case of the Brown & Williamson Tobacco Corporation versus Merrell Williams.

Merrell Williams worked as a paralegal for a law firm hired by Brown & Williamson to review documents in preparation for lawsuits that had been brought against B & W. Documents on B & W's internal discussions in the 1960s and 70s were leaked to members of Congress and *The New York Times*. B & W named Williams as the culprit and filed suit against him. In January of 1995 Judge Thomas Wine of the Jefferson County Circuit Court in Louisville, Kentucky, issued an order barring Williams from disclosing, using, or reproducing any stolen documents and information in his possession. Judge Wine's decision was unanimously upheld by the Kentucky Supreme Court in February, which meant that Williams could face criminal charges without being allowed legal counsel on these matters. This raised questions about whatever Williams' constitutional rights were being violated. A spokesperson for B & W said that it was pleased with this ruling. Not for too long, however; in April Judge Wine reversed his original temporary injunction noting that in considering the extensive public access the documents had received, it would be unconscionable to deny William effective representation.

Why hound Williams to ground? Revenge might be sweet, but much more important is to send a message through Williams to other potential leakers; cross us and this is a sample of what you're in for.

2. Forestall Product Liability Suits. Product liability suits are troublesome for three reasons: The process of defending against such is expensive; settlements or judgments might run into a considerable amount of money, millions even billions of dollars; and, potentially most deadly, damaging disclosure. The lawsuit filed by Ilo Grundberg against the Upjohn Company forced Upjohn to make available a good deal of data about its sleeping pill Halcion. Upjohn thought it wise to limit its losses by settling the suit, but a good deal of unfavorable publicity was still generated. General Motors was forced into court by the Moseley family over the death of their son Shannon who was killed when his G.M.C. pickup truck was struck on the side and burst into flames; a public relations nightmare ensued. Manufacturers of silicone breast implants thought it wise to settle a class action suit for $4.2 billion.

A nationwide class action suit against the tobacco industry is in preparation and some states are also gearing up to file class action suits on behalf of Medicaid patients who smoke or have smoked. Such developments, needless to say, send chills down the collective spine of Corporate America.

What can be done? Simple; change the product liability law to make it very difficult to file suit. Although cloaked in terms of saving the consumer money which Corporate America must pay to rapacious lawyers to defend against suits, this is one of the objectives of Representative Newt Gingrich's Contract with America.

3. Develop "Plausible Deniability." Admiral John Poindexter sought to do this for President Reagan in his testimony on Iran-Contra. The tobacco industry has sought to do this in connection with mounting scientific evidence that nicotine is addictive and smoking is hazardous to health. It is a versatile tool which can be packaged in many ways.

4. Seek Cloaks of Respectability. If you are on the board of a tobacco company, seek membership on the board of a prestigious health establishment, Memorial Sloan-Kettering Cancer Center, for example. Would they not consider this a conflict of interest and refuse to accept you? Not necessarily. (For an interesting correspondence on this subject, see Edward T. Chase[6]). Some would suggest

that there is a natural concurrence of interests in such memberships in that one institution helps provide the clients for the other to cure.

5. Muffle Negatives by Emphasizing Positives. This involves the careful selection of buzz words which the public has come to react to positively or negatively. For example, to counteract a thrust to regulate the amount of nicotine in cigarettes and smoking in public places, emphasize the individual's freedom of choice as being threatened by big brother government. To make palatable restrictions on product liability suits, emphasize the savings that accrue to consumers of money that would otherwise have to be paid to rapacious lawyers to defend against such suits.

In this setting lawyers are not just lawyers, which lacks sufficient color. They are to be described as rapacious or greedy lawyers. When you need one for a positive point being made, they are to become knights or defenders of law and order or protectors of the rights of the people from the encroachments of big government, or big whatever.

CAUTION, SPEED, RELIABILITY: WE WANT IT ALL

4

WE NEED THE DATA, NOW

Situations arise, particularly when lives are at stake, when there is an urgent need for data. But the data must also be reliable, which calls for the exercise of caution. The undertaking requires time, an ingredient which cannot be rushed without running the risk of putting in jeopardy the success of the study and the lives that it seeks to save.

THE DRUG THAT ALMOST MADE IT

On September 4, 1991 an advisory panel of the Food and Drug Administration recommended that the experimental drug centoxin be approved to treat sepsis, a reaction to a bacterial toxin in the bloodstream that kills an estimated 100,000 people a year in the United States. Centoxin was expected to save thousands of lives and bring a high profit to Centocor, Inc., the company that developed it. Clear sailing to final FDA approval seemed certain, but then the storm broke.

Dr. Jean-Daniel Baumgartner, a Swiss scientist who undertook to test centoxin for the French company Merieux, was unable to reproduce the data from laboratory and animal studies used by Centocor to justify testing the drug in humans. One of the F.D.A.'s reviewers of Centocor's license application noted that the company had changed its criteria for success while the initial study of the drug was underway.

Centocor was ordered to conduct another study, but in early 1993 it halted the second study and withdrew the drug from the ten European countries where it had been approved. In the study, one group of patients with sepsis showed a higher death rate among those who had taken the drug than among those who had not taken it. Some doctors who have worked with centoxin believe that it works, provided that you can find the "appropriate" patients. Doctors also point out that had the drug been approved, it would not have been possible to tell how effective it is. Patients who received the drug and lived might have done so anyway and those who died might have done so because the drug was ineffective.

The centoxin experience illustrates, medical authorities argue, why premature licensing of a drug because people are dying and there is a chance that it might work might do a good deal of harm.

THIS DRUG MADE IT, BUT QUESTIONS REMAIN

The standard early treatment prescribed for those infected with the virus H.I.V. that causes AIDS is the drug AZT.

In early 1993 a study carried out in England, France and Ireland, called the Concorde study, found that there is no clear benefit in taking AZT early. Two groups of H.I.V. infected patients were compared. For one group the drug treatment was begun during the symptom free period of the disease and for the other it was withheld unless symptoms appeared. There was no "real difference" in the rates of progression to AIDS or death, the study found.

The findings differed from those of four smaller and shorter studies in the United States. Three of these studies were stopped ahead of schedule for "ethical reasons" because those who had received early treatment seemed to benefit. The patients were followed for a year or less in three of the studies and two years in the fourth. The Concorde study followed patients for an average of three years.

A study conducted in Australia and published in the summer of 1993 concluded that AZT slows the progression to AIDS. Since the

Concorde and Australian studies used different measuring sticks for determining disease progression, different follow-up periods and end points, direct comparison of the conclusions reached are at best "difficult."

Which studies are the reliable indicators? Not clear.

RUNNING THE STOP SIGNS

Scientific studies have a methodology of their own and the adoption of short-cuts to get the data or win approval for drug sales is fraught with pitfalls that can lead to counterproductive or disastrous results. Centoxin, which had been recommended for approval by an advisory panel of the Food and Drug Administration, is a case in point. An eleventh hour reprieve saved the public from a potential disaster.

Another episode which illustrates what may happen when the stop signs are ignored involves the development of a multiple drug treatment for AIDS based on the strategy of employing a combination of the drugs AZT, dd1, and nevirapine to retard the development of AIDS. At a news conference held in February of 1993 Yung-Kang Chow, a Harvard medical student, announced that the team of which he was part had found the "Achilles heel" of H.I.V.

The reaction and ensuing publicity were enormous. The "Achilles heel" was a theory about H.I.V. structure that justified the multiple drug test treatment for AIDS. The preliminary data were ambiguous. Chow rushed into publication without repeating the crucial steps to verify the data and, pushed by pressures to attack AIDS with every potential resource available, Federal health officials authorized national clinical trials involving 200 participants without independent confirmation of the research findings. In July the bubble burst and the Harvard team announced that its research was fatally flawed.

The DES and thalidomide stories illustrate drugs whose effects reached beyond their takers to the next generation. DES, short for diethylstilbestrol, is a synthetic hormone which mimics the natural estrogen hormone estradiol. It was prescribed to approximately five million pregnant women between 1947 and 1971 for threatened miscarriage. The Food and Drug Administration barred pregnant women from using DES in 1971 when a link between women who

took the drug and rare cervical and vaginal cancers developed by their daughters was found. DES has also been linked to breast cancer in the mothers, infertility and other reproductive problems in their daughters, and possible testicular cancers in their sons. Some cancer sufferers have won jury awards and in January 1994 a jury awarded $42.3 million to eleven women whose mothers took DES during pregnancy, including eight who claimed that it caused reproductive problems not related to cancer.

DES is currently used to treat symptoms of underdeveloped ovaries, menopause, and atrophic vaginitis.

The thalidomide story, which goes back to the early 1960s, has an instructive but bittersweet ending. Thalidomide is a sleeping drug which caused limb deformities in an estimated 10,000 babies born to women who took it during pregnancy. This occurred in Europe, Australia, and Canada where the drug was available by prescription and even sold in some countries, such as Germany, without prescription. It was not approved in America because of the caution exercised by Dr. Frances Kelsey of the Food and Drug Administration who refused to approve the Merrell Company's request to market thalidomide here because she believed that its safety had not been thoroughly investigated. Before thalidomide's link to limb deformities became apparent, she was denounced by some as a bureaucratic nitpicker who was keeping a "miracle drug," as thalidomide had been described, from the country. Subsequently, she was hailed as a hero.

The problem of determining when a new drug is ready for the market is one of society's great challenges. If government regulations are "unreasonably" rigid in demanding proof of a drug's effectiveness and safety, patients may die before having a chance to benefit from it. If the testing of a drug is moved to the fast track and it is approved prematurely, patients and their offspring as well might suffer harm and even death, and millions of dollars may be wasted by those who pay the bills.

CAN LESS BE MORE?

The Food and Drug Administration is often treated like a not-so-favored Rock group whose audience wants it to sing less. Pharmaceutical and biotechnology companies favor a fast track for the approval of drugs and biotechnology products; time, after all, is money and the process of obtaining reliable data may be time consuming. Advocacy groups campaign for a fast track for the approval of particular drugs; time, we are told, may mean life. And then there

are groups that, under the banner of conserving taxpayer dollars, would reduce the funding of the F.D.A. or, under the banner that chickens are best guarded by foxes, would turn the functioning of the F.D.A. over to private industry. There are also those who advocate allowing drugs to be marketed without F.D.A. approval but carry a warning that they have not been F.D.A. approved—the best guinea pigs are in the public itself approach.

There is no question that lives, time, and cost are legitimate concerns, but experience has made clear that beyond them stands the unyielding fact that there is no substitute for sound data. Precipitous shortcuts can be extremely costly in terms of the lives, time, and money they may have been intended to save.

WHAT ARE THE NUMBERS AND WHAT DO THEY TELL US?

"GOOD" NUMBERS WOULD HELP

Numerically stated conclusions without a source or clear basis often do not stand up under close scrutiny. Neither do bold assertions with numerical implications.

The previous cases make clear that when figures are presented one would be well advised to be on guard. The following cases make clear that when figures are not presented, contradictory figures are pre-

sented, or illusionary figures are presented, one should be even more on guard.

WHY THE JOB LOSS?

Karen-Alicia Robertson claims[1] that four of every ten dollars that Oregonians earn are made from the harvesting and cutting of trees. As is often the case with advocacy type letters, no source was given for this figure. Its suggested inference is that if the remaining natural forest of old-growth trees is preserved, Oregon faces massive unemployment, perhaps reaching 40 percent of its jobs.

Frederic Sutherland, in a reply[2], called into question Robertson's statistics and noted that employment in Oregon's timber industry had fallen by 20 percent in the previous ten years, while the amount of timber cut in the state's national forests had risen by 18 percent. Sutherland gives a source for his statistics.

NUMBER MAGIC

In early 1981 Ronald Reagan's budget director David Stockman found himself drowning in a lake of red ink. President Reagan had promised that the country would have a balanced budget by 1983, and Stockman found himself forced to move the target date to 1984.

Stockman writes:[3]

> But that was merely a straw in the wind compared to what would come next. I soon became a veritable incubator of short-cuts, schemes, and devices to overcome the truth now upon us—that the budget gap couldn't be closed except by a dictator.

> The more I flopped and staggered around, however, the more they went along. I could have been wearing a sandwich board sign saying: Stop me, I'm dangerous! Even then they might not have done so . . .

> Bookkeeping invention thus began its wondrous works. We invented the 'magic asterisk': If we couldn't find the saving in

time—and we couldn't—we would issue an IOU. We would call it 'Future savings to be identified.'

It was marvelously creative. A magic asterisk item would cost negative $30 billion . . . $40 billion . . . whatever it took to get a balanced budget in 1984 after we toted up all the individual budget cuts we'd actually approved.

The magic asterisk passed presidential and congressional muster.

DOES THIS SCENARIO ADD UP?

In a speech given on January 3, 1992, Representative Newt Gingrich of Georgia took aim at the "welfare state." Citing the case of an Atlanta drug addict who, while on parole, had been accused of killing members of his family, he said that the nation should build enough prisons so that there are enough beds that every violent criminal in America is locked up. He noted that we could adopt three or four military bases for this purpose and that construction could be carried out at bargain rates by National Guard engineering battalions.

Putting some numbers to this scene yields a cost dimension which calls its viability into question from a cost point of view alone. The number of violent crimes committed each year exceeds five million and the construction cost of a single cell is close to $100,000, irrespective of who does the work. And then there are the costs of maintenance and staff.

THERE'S MORE HERE THAN MEETS THE EYE

Prior to the 1992 election, President George Bush and Secretary of Education Lamar Alexander were hard at work promoting education reform based on the voucher system. A G.I. Bill for kids was envisioned as providing $1000 for each of 500,000 children from low and middle income families in fifty communities to help them pay for a private school education. Envisioned cost: $500 million.

If one considers the approximately 5 million students attending private schools with voting parents, friends and supporters who would strongly feel that equity requires that they be granted such relief, or no votes, the cost is brought to a total in the neighborhood of $5.5 billion.

With expectations having been aroused, suppose that a relatively small number of additional families with children currently in public schools say that they want their children to attend private schools with the government paying at least part of the bill. The political pressure to accommodate an additional 1 to 2 million students in the voucher system could be enormous. If successful, this could add an additional $1 billion to $2 billion to the bill.

SILENCE MAY BE GOLDEN AFTER ALL

Friends and foes alike of Senator Lloyd L. Wind, who was famous for his lengthy and somewhat incoherent oratory before that August body, decided that they needed a break, as modern parlance would put it. A bipartisan committee was authorized to offer "Long," as he had come to be called by all, the following deal. "Beginning next Monday we will donate one dollar to the charity you designate and double the size of the donation every week thereafter for as many weeks as you remain silent in the Senate." "Done deal," said Long, "put it in writing."

Those represented by the bipartisan committee underestimated how rapidly money can grow under the conditions of the deal they offered and "Long" Wind's willpower to stay the course when the right incentive is provided. This was unquestionably the right incentive. Twenty weeks later Senator Wind's colleagues emerged from a

well earned rest from the good Senator's oratory for which they paid his favorite charity $2,097,151.[4] Despite the unanticipated cost, the final consensus was that it was worth it.

LET'S GET THE DATA

This is not a case of the data not being available; they are. But they are outdated, which makes them illusionary. In a sense one is better off not having the data at all; then it is unequivocally clear where one stands.

In the fall of 1993 actuaries from several large accounting companies concluded that White House estimates of health insurance costs for families and individuals were based on outdated data. The latest comprehensive National Medical Expenditures Survey, for example, goes back to 1987 and does not reflect recent changes. The latest detailed Government survey of state health spending goes back to 1982.

Realistic health cost figures lie at the foundation of any viable health insurance plan.

LET'S GET THE DATA 2

A report issued in the fall of 1991 by a scientific panel and the General Accounting Office criticized the latest National Food Consumption Survey, carried out in 1987–88, as so flawed that its data are probably useless. The major problem is the survey's low response rate of 34 percent, making it questionable whether the data are representative of the population. Follow up studies are required of those who do not respond, but no follow up studies were conducted.

The flawed data are used for making major Government policy decisions involving school breakfast and lunch programs, food stamp allotments, setting pesticide levels in foods, calculating nutrient consumption levels, and determining the public's exposure to pesticides and toxic metals.

LET'S GET THE DATA 3

In November 1993 the General Accounting Office reported that the nutrition data in the publication Handbook 8 is flawed because of sloppy, inconsistent, or questionable methods of collecting data. Handbook 8 states, for example, that there are 3000 international units of Vitamin A in a portion of papaya while it is known that there are 400 units in the same portion. Nutrition data on bacon-cheese-burgers, to take another example, comes from brochures provided by fast food chains, which do not explain the basis for their claims.

Handbook 8 is a major resource of nutrition data. It plays a role in determining public nutrition policy, planning feeding programs, medical research, and providing information for folks on diets who are trying to keep track of such things as calories, fat levels, and sodium intake.

STATISTICS MAY NOT LIE, BUT WHAT DO THEY SAY?

WHICH INTERPRETATION IS "CORRECT"?

How statistics with the limitations we have imposed are interpreted will, at best, depend on the judgment and capacity of the interpreters and, at worst, on the cleverness of the spin doctors entrusted with putting on them the best possible face. It is often the case that "experts" see very different things in the same statistics and that different statistics concerning the same issue seem to be contradictory.

THERE'S LESS TO BASEBALL STATISTICS THAN MEETS THE BAT

In his book *The Last Yankee: The Turbulent Life of Billy Martin*, David Falkner[1] concludes that Martin was the best manager of his era, possibly of many eras. Falkner's judgment was strongly influenced by baseball statistics compiled by the Elias Sports Bureau and a formula which claims to show which managers' teams won more games than they were reasonably expected to win.

In his review of Falkner's book, George F. Will[2] disputes Falkner's conclusion which, he argues, the rest of the book refutes. "In fact," Will notes:

> *The Last Yankee* might usefully be made required reading for graduate students in the social sciences and all others who need to be immunized against the seduction of numbers. . . .

There are limits—and Mr. Falkner's reporting shows that Elias passed them regarding Martin—to the ability to capture messy reality in tidy formulas.

HOW SUCCESSFUL WAS THE PATRIOT?

During the Gulf War, television viewers were moved by scenes of Patriot missiles streaking across the sky to intercept Iraqi Scud missiles that had been launched against Israel and Saudi Arabia. The Patriot's success seemed to epitomize the success of a high tech, low causality military campaign.

Apart from very successful military public relations, how successful was the Patriot as a military tool? Different statistics have been

given and, as is almost always the case, it is important to look further than the statistics if a realistic picture is to emerge.

The Army originally stated that Patriots "intercepted" 45 of 47 incoming Scud missiles, and President Bush revised that to 41 of 42. What does this mean? Brigadier General Robert Drolet of the Army's Missile Command testified that "a Patriot and a Scud passed in the sky." There are other statistics of interest. Before Patriots were employed in Israel, 13 Scuds fell near Tel Aviv. There were no deaths, but 115 people were wounded and 2,698 apartments were damaged. After Patriots were employed in this region, 11 Scud attacks left 1 dead, 168 injured and 7,778 apartments damaged.[3] This is explained by the fact that successful hits led to more deadly debris being sprayed over a larger area than otherwise would have been the case and that the Patriots tended to strike the bodies of Scuds, leaving their warheads armed and able to cause significant damage on landing.

POLITICAL FIGURING

Rattled by a setback to Jerry Brown in Connecticut in his quest for the Democratic Presidential nomination, Bill Clinton and his campaign supporters stormed into New York determined to make his nomination seem inevitable again. The primary held on 7 April 1992 left Clinton with 41 percent of the vote to Paul Tsongas's 29 percent and Jerry Brown's 26 percent; former Senator Paul Tsongas had earlier suspended his campaign.

Clinton's 41 percent clearly outdistanced his rivals, and Clinton and his aides claimed that this victory was a reflection of the strength of his message, which helped voters overcome doubts that may have been raised by persistent attacks on his character. The same statistic made clear that almost 60 percent of the voters preferred another candidate, with 29 percent opting for a candidate who had withdrawn from the race. Other statistics reflected a less than enthusiastic endorsement of Clinton's candidacy with implications for the Presidential race itself. Just 27 percent of the registered Democrats turned out to vote, the lowest turnout in years. The exit poll conducted by voters Research and Surveys showed that nearly two-thirds of the voters would like to see other candidates enter the race and that almost half had reservations about Clinton's honesty and integrity.

TEST SCORES RISE: EDUCATION MIRACLE OR MIRAGE?

In an era marked by bleak news on the education front with school systems finding themselves more and more driven to distraction, a June 1995 announcement that math and reading scores of elementary school students rose in all 32 New York City school districts that spring was greeted with pride and optimism that the City had turned an education corner. School officials attributed the rise to a new curriculum framework, improved guidelines for what children should be learning and when, and extra help for teachers in the City's 100 lowest performing schools.

But is there as much here as meets the eye? To what extent can an increase (or decrease, for that matter) in test scores be attributed to

mastery of subject matter and to what extent can it be attributed to mastery of the measuring instrument—the exam itself? With standard exams that are continually being administered, grades tend to improve as teachers master the exam format and become adept at running practice sessions. (This is called teaching to the exam in those extreme cases where preoccupation with the exam overrides preoccupation with the subject itself.) Grades tend to decline when there is an abrupt change in the exam format which teachers and students have not had time to adopt to. (Such exams are often called "unfair.")

THE REAGAN ECONOMIC BOOM: BLESSING OR DISASTER?

Martin Anderson, former advisor to President Reagan and senior fellow at the Hoover Institution, employs statistics to support his view that the Reagan economic boom was the greatest ever:[4]

The two key measures that mark a depression or expansion are jobs and production. Let's look at the records that were set.

Creation of jobs.

From November 1982, when President Ronald Reagan's new economic program was beginning to take effect, to November 1989, 18.7 million new jobs were created. It was a world record: . . . The new jobs covered the entire spectrum of work, and more than half of them paid more than $20,000 a year. As total employment grew to 119.5 million, the rate of unemployment fell to slightly over 5 percent, the lowest level in 15 years.

Creation of wealth.

The amount of wealth produced during this seven year period was stupendous—some $30 trillion worth of goods and services. Again, it was a world record. . . . According to a recent study, net asset values—including stocks, bonds and real estate—went up by more than $5 trillion between 1982 and 1989, and increase of roughly 50 percent. . . .

Income tax rates, interest rates and inflation.

Under President Reagan, top personal income tax rates were lowered dramatically from 70 percent to 28 percent. This policy change was the prime force behind the record breaking economic expansion. . . .

The stock market.

Perhaps the key indicator of an economy's booms and busts is the stock market, the bottom line economic report card. . . . starting in late 1982, just as Reaganomics began to work, the stock market took off like a giant skyrocket. Since then, the Standard & Poor's index has soared, reaching a record high of 360, almost triple what it was in 1982.

There were other consequences of the expansion. Annual Federal spending on public housing and welfare, and on Social Security, Medicare and health all increased by billions of dollars. The poverty rate has fallen steadily since 1983.

When you add up the record of the Reagan years, and the first year of President Bush. . . . the conclusion is clear, inescapable and stunning. We have just witnessed America's Great Expansion.

In a reply, Nobel Prize winning economist Wassily Leontief[5] concedes some of Anderson's statistics but goes on to look at a number of cost thorns in his statistical rose garden.

True, the long recovery from the deep depression that brought President Reagan to power carried this country to the high point of the usual cyclical wave characterized by a low rate of unemployment and a high gross national product. It is most likely that wholesale tax cuts inaugurated by Mr. Reagan have made the level of the G.N.P., as measured by the Government statisticians, several billion dollars higher than it would otherwise have been. But at what a cost!

Drastic cuts in public spending (except for military purposes) left the physical infrastructure of this country in ruin. City streets and transportation facilities, water supply and sewage systems, particularly in large metropolitan areas, are collapsing, the once glorious interstate highways are crumbling, and cramped airports are incapable of handling the rapidly increasing traffic. Despite the valiant effort of the underfinanced, underpowered Environmental Protection Agency, our lakes, rivers and forests are succumbing to deadly acid rain.

What is even worse, the intellectual, cultural and social infrastructure of the country has suffered even more during this greater-than-ever boom than its physical counterpart. Primary and secondary schooling have been so weakened that a whole generation of boys and girls can hardly read, write or count, while the soaring price of higher education makes it impossible for many young people to take advantage of it.

No wonder the competitiveness of the United States is rapidly declining; many of our high technology industries are losing one battle after another in the struggle for their share of the foreign and even their own domestic market. At the same time, the rich are getting richer, and the poor are getting homeless.

Let us hope that contrary to Mr. Anderson's expectations the "Reagan Boom" will not continue in its present form for four or eight more years. If it does, the United States will find itself entering the 21st century as the richest country (in total value of stocks and bonds traded on the stock exchanges), but culturally and socially less advanced than other developed countries.

MORE SPIN AND COUNTERSPIN

In the first three months of 1993 the economy grew at the slow pace of 1.8 percent, Commerce Department figures showed. What does it mean?

According to President Clinton, who was trying to get a Government economic stimulus package through Congress, this figure made very clear the need for his stimulus package which was being blocked by Senate Republicans. "I told you so," was the reaction of his Administration.

"The American people are grading the President with their pocket books," thundered Senator Bob Dole, the Senate minority leader, in reply. Senator John Chafee from Rhode Island argued that "what these figures reflect is that people who are going to spend money . . . are scared of the mammoth new taxes that have been proposed."

Herbert Stein, chief economic advisor to President Nixon, noted that "it doesn't affect my basic outlook that we'll have a moderate recovery." Charles Schulze, chief economic advisor to President Carter, commented that the figures were "consistent with nonbooming, moderate" expansion.

IF YOU DON'T AGREE, SHUT UP!

In 1986 a team led by Dr. Charles Bluestone submitted a paper to the *New England Journal of Medicine* which concluded that the antibiotic amoxacillin was effective in treating middle-ear infection. Dr. Erdem Cantekin, a member of the team, dissented from this conclusion and submitted a separate report to the *Journal* which he

claimed was a re-analysis and re-interpretation of the paper that Bluestone had submitted a month earlier.

The *Journal's* editor saw the issue as determining which group had the right to publish and, on the advice of officials at the University of Pittsburgh and the hospital where many patients in the study were treated, decided to publish the Bluestone paper only. The papers were not sent out for review. Shortly after submitting his paper for publication in 1986, Dr. Cantekin was removed as director of Pittsburgh University's center for studies of middle-ear infections.

Interests, egos, and procedures had collided to the detriment of open inquiry in medical science. However, Dr. Cantekin persisted, and saw his report published in *The Journal of the American Medical Association* in December 1991.

DOES BALDNESS CAUSE HEART ATTACKS?

A study of men under 55 published by *The Journal of the American Medical Association* in February 1993 concluded that baldness on the top of the head, called vertex baldness, is correlated with a slight but definite increase of heart attack. The more extensive the vertex baldness, the higher the risk, the study found.

Does this mean that vertex baldness causes heart attacks? NO! What has been found is a statistical link between two factors. A statistical link between two factors does not imply cause and effect. It is possible that some third factor influenced either or both variables or that the correlation is spurious.

There is, for example, a well known relationship between the number of beds in a region and the death rate. This obviously does not mean that lying in bed is a cause of death. A bed is where many find themselves when they pass on. There is a statistical correlation between Scholastic Aptitude Test score and academic success to the extent that SAT score is used by many colleges as part of their admissions profile. The SAT score is not a cause of academic success. At best there are behind-the-scene characteristics that influence the SAT score and academic success in such a way that one might be used as a "successful" prediction tool for the other.

Before making a major change in life style, what can we do to weigh the evidence in an age when alarming reports of health risks connected with almost every aspect of life have become commonplace? The following guidelines are useful, but only as guidelines.

1. Keep in mind that statistical correlation does not imply cause.

2. Consider the source of the report and the publisher. There are no guarantees, but it is reasonable to treat a study done at a major university, such as Harvard or Yale, as more authoritative than one coming from an unknown school. Reports published by peer-reviewed journals, such as *The New England Journal of Medicine*, are more reliable than those that come out of medical meetings or news conferences. An editorial should accompany major studies discussing their strengths and weaknesses.

3. Are there other studies which have reached the same conclusion?

4. Is the finding supported by animal studies?

5. Are there biological explanations for the observed association?

WEIGHING THE DATA: HOW CREDIBLE IS CREDIBLE?

We return to the silicone breast implant case (see *Women's Nightmares,* Ch. 3), a case of vital significance in its own right, but which also raises important general questions about weighing data and setting public policy. These are the questions: What does the data say about the safety of silicone breast implants? What are the implications for the more than one million women who have had breast implant surgery and others who may be considering this procedure? What are the implications for legal actions and public policy?

The following developments are noteworthy.

Early 1960s

Silicone breast implants begin to appear on the market. At this time there were no Federal regulations that required proof of safety for medical devices. Implant makers were not required to make a case that they were not unsafe. Federal regulation of medical devices began in 1976.

1970

The implant study involving dogs noted in *Women's Nightmares* is undertaken.

1975

A study conducted by Dow Corning shows that a form of silicone gel called D4 is highly toxic to the immune system of mice when used in purified form. It is the first study to show that a specific silicone is harmful to the immune system of mammals. The study was not reported to the Food and Drug Administration (F.D.A). Its existence was discovered in April of 1994 by lawyers representing breast implant patients in a class action suit.

1983

An internal memo by Dow Corning scientist William Boley notes: We have no valid long-term data to substantiate the safety of gels.

1991

The first wave of lawsuits by women with implants who suffered from a variety of autoimmune diseases and problems descends on implant manufacturers. Multimillion dollar judgments for plaintiffs whose cases went to trial begin to appear.

November: On reviewing evidence submitted by a number of silicone breast implant manufacturers, a panel of F.D.A. advisers concludes that none know whether they are safe. The panel recommends that the product be left on the market while the manufacturers conduct safety studies.

1992

January: F.D.A. Commissioner Dr. David Kessler asks manufacturers and physicians to voluntarily discontinue the sale and use of silicone breast implants. The moratorium is temporary, pending further review.

April: The F.D.A. approves limited use of silicone breast implants, subject to the women being monitored in clinical studies.

1993

Early: Dr. John Naim of Rochester General Hospital shows that pulverized silicone gel from breast implants injected into rats has a strong effect on their immune systems. Naim's study was replicated by Dow Corning scientists.

March: In examining the blood of women with breast implants who came to the Medical Center at the University of California at Davis School of Medicine, researchers find that 35 of 100 women examined had antibodies against their own collagen, which meant that their immune systems were reacting to their own bodies. With lupus and rheumatoid arthritis, for example, the body's immune system mistakes its own tissue for foreign material and attacks it. The California finding suggests that some women with silicone gel implants may be experiencing such a mistaken immune system reaction.

November: The American Medical Association (A.M.A.) expresses the view that until additional health hazards were established, silicone breast implants should be available without restriction to women who were fully informed about the risks and benefits. (*The Journal of the American Medical Association*, Nov. 30).

1994

March: Implant manufacturers and plaintiffs agree on a settlement that will pay $4 billion over 30 years to compensate women with implants who file claims based on diagnoses of any one of several conditions. The companies continue to insist that implants are safe, but that they have agreed to the settlement to avoid protracted litigation.

June: An epidemiological study conducted by Dr. Sherine Gabriel of the Majo Clinic reports that 749 women living in Olmsted County in southern Minnesota who had implants from 1964 to 1991 exhibited no obvious excess medical problems.

This was the third epidemiological study to find no excess medical problems among women with implants as opposed to the general population.

In an editorial accompanying Dr. Gabriel's study, Dr. Marcia Angell, Executive Editor of the *New England Journal of Medicine*, again criticized the F.D.A.'s decision two years earlier to impose a moratorium on the use of silicone gel implants, an action which she viewed as being taken, not because the implants had been proved harmful, but because manufacturers had not done the legally required studies to establish that they are safe. She and others view the results of the epidemiological studies as confirmation that the F.D.A. had acted precipitously.

WHO HAS THE BURDEN OF "PROOF"? WHICH DATA ARE CREDIBLE? Dr. Angell's comment raises a fundamental question: Does the manufacturer of a medical device have the primary responsibility of proving that its product is not harmful once it has been allowed on the market and questions have been raised, or does the F.D.A. have the primary responsibility of proving that the product is harmful before it is allowed to continue on the market? Also, what constitutes "credible proof"? If the first priority of the F.D.A. is to protect the public from harm, then the focus of proof must be to secure evidence that use of a product does not run the risk of "serious" side effects. This burden of proof falls to the product's manufacturers, a burden which silicone gel manufacturers were unable to sustain. They assumed that silicones are not biologically active and are thus suitable materials for breast implants. Research, some done by Dow Corning as early as 1975, has not supported this view. Considering the research evidence available, the question is not whether silicones are biologically active, but how active and with what consequences. From this point of view, the F.D.A. had no choice but to, at the very least, request a moratorium on the sale and use of silicone gel implants. Elsewhere, the French Government withdrew silicone breast implants from France and ordered their manufacture, distribution, and importation stopped. The decree of May 17, 1995 states that France will not allow silicone gel implants back on the market until they have been shown to be without risk.

Did the Gabriel and two other epidemiological studies vindicate Dr. Angell in her judgment that the F.D.A. had acted precipitously? Alas, it is not that simple. Epidemiological studies examine a group of people in a search for a statistical pattern. If a pattern is found, this does not by itself establish cause and effect. If a pattern is not found, this does not by itself show that there is no relationship between the underlying factors. Epidemiological studies should not be confused with controlled statistical studies in which a number of subjects with a factor of interest are chosen and observed and a number of similar subjects without the factor of interest are chosen and observed. With epidemiological studies we are constrained by the number of subjects on the spot and we have no control over possible confounding factors. If the number of subjects observed is relatively small and no pattern is discerned, questions about whether a larger scale study would be more revealing and whether confounding factors obscuring a pattern were at work arise. Epidemiological studies on silicone gel breast implants carried out to date add to our information picture, but are not decisive. They raise more questions than they answer.

CREDIBLE PROOF IN LITIGATION. To make sense of why major silicone gel breast implant makers were willing to agree to a $4 billion settlement in a class action suit when they felt there was no scientific evidence that silicone gel implants cause serious disease, it is important to keep in mind that standards of credible scientific proof (the phrase beyond a reasonable doubt in a scientific sense comes to mind here) and credible legal proof for a plaintiff to prevail in a civil suit (preponderance of the evidence) differ. As plaintiffs began to win multi-million dollar judgments based on a preponderance of the evidence standard, implant makers thought it wise to limit their loses by negotiating a settlement.

The $4 billion settlement began to unravel when, faced with the possibility of losses that might substantially exceed their share of this amount, Dow Corning filed for Chapter 11 bankruptcy protection on May 15, 1995.

7 ► ARE WE ON THE RIGHT DATA TRAIL?

RELIABILITY VERSUS RELEVANCE

A situation requiring study arises and word goes out to "get the data." What data should be sought? If our concern is to not exceed the weight limitation of cargo to be shipped, data about the volume of the cargo, no matter how reliable it might be, is not relevant to this concern.

Determining which data are relevant to a study being undertaken is often not a clear cut matter. A poorly formulated framework for the study may lead to data that are at best partially relevant to the objective of the study, although reliable in their own right.

Data themselves do not lie, but what they give us cannot exceed the rules and limitations we impose to define the framework from which they come. If that framework is overly simplistic, then so too will be the data which it yields.

A VISIT TO HUXLEY COLLEGE

Huxley College cannot be located on a map, but its spirit and problems are as real as those colleges and universities which can be so determined.

Ivor M. Wisdom, a financial analyst, was hired by President Marx of Huxley College to analyze the operations of the departments of Huxley College and make recommendations on how to improve their financial efficiency. I.M. Wisdom defined the income of each department as the tuition income of the students being serviced by the department minus costs, primarily salary costs. He collected data on the class size of each instructor and each instructor's rank and salary and found that a number of full professors at the top of the salary scale were teaching classes with a small number of students. To improve the income of the departments, Wisdom recommended that teachers at the top of the salary scale be assigned basic level courses which can be expected to have a large number of students.

There is no question that Wisdom's data were consistent with his view of financial efficiency and reliable, but are they relevant to the issue? "Horsefeathers," cried Dean Marx, President Marx's brother. "It's irrelevant to the overall financial efficiency of the departments and Huxley as a whole. You could change the teaching assignments at the last minute and this game of academic musical chairs will not change a department's overall tuition revenue or costs on which financial efficiency depends. The scheme is counterproductive in that it deflects us from the real issue of financial efficiency. It may also have negative academic consequences if taken seriously by

leading us to assign courses to be taught on the basis of an instructor's spurious personal efficiency rather than academic qualifications."

WHICH DATA "BEST" REFLECT AIRLINE RELIABILITY?

The long time standard measure of an airline's reliability is its percentage of on-time arrivals, where a flight is deemed on-time if it arrives within 15 minutes of its scheduled arrival time. Such data are widely trumpeted by airlines in their advertising campaigns.

But is this statistic the "best" measure of an airline's reliability? According to Julius Maldutis, an airline analyst with Salomon Brothers, the answer is no. Maldutis argues that a much better measure of reliability is the percentage of flight-miles completed. Look at the cancellation rate, which is indicative of a more troublesome situation to travelers than that indicated by the artificial on-time statistic, says Maldutis.[1]

THE NATIONAL DEBT: PUBLIC NUISANCE OR MENACE?

By the beginning of 1993 the gross national debt of the United States, it was generally agreed, was in the neighborhood of $4.2 trillion, a staggering figure which boggles the mind. If we think of one dollar as equivalent to one second, $4.2 trillion takes us back about 130,000 years to the Middle Pleistocene period in which the pre-neandertaloids are found. If you had to transfer this amount of money in $100 bills from one location to another, you would have to deal with a stack of bills 2670 miles long.

The figure sounds ominous, but here is where disagreement begins. One point of view argues that the figure itself and the rate at which it has been increasing portend catastrophic consequences in the offing. When the debt grows faster than the country's ability to carry it, a breakdown with social, political, and economic upheaval is inevitable, and we are coming dangerously close to this state, this view has it. Its proponents include former Senators Paul Tsongas and Warren Rudman who formed the Concord Coalition to rally public support for making deficit reduction a top priority, James Davidson, founder of the National Taxpayer's Union, and Ross Perot, who made debt reduction the major theme of his 1992 presidential campaign.

Another view argues that in terms of the state of the economy, we are looking at the wrong figure and that, while debt reduction is desirable, it should not be given top priority and carried out in a "mindless" way since this will severely damage the economy. Its proponents focus on the ratio of publicly held debt to Gross Domestic Product (GDP), now 51 percent, which they argue is not a figure to be alarmed over. The major problem is the low productivity growth rate, this view argues, and that budget cutting will make it impossible to do what had to be done in the way of education and infrastructure repair to enhance productivity. The problem is not to cut the budget to reduce the deficit but to reallocate the budget away from consumption to investment that will increase productivity. This camp includes Lester Thurow, formerly Dean of the Sloan School of Management of M.I.T., and Robert Eisner, professor of Economics at Northwestern University and past president of the American Economic Association. Among many writings on the subject, see Notes 2–4 for a more complete discussion of these points of view.

IS THE RECESSION OVER IF THE STATISTICS SAY SO?

In early 1992 President Bush was, as modern parlance would put it, an unhappy camper. Government statistics showed a mild recession and strong economic fundamentals. Yet business and consumer confidence in the economy had been shaken to an extent that seemed way out of proportion to the statistical signs, and many were blaming George Bush for having missed the wake up call.

"The problem," observes Charles McMillion,[5] "is that many of those statistics are wildly misleading."

One statistic concerns unemployment. During the 1982 recession, the worst since World War II, unemployment reached 10.8 percent. During the 1991–92 recession, it reached 7.8 percent. McMillion notes that the 1991–92 unemployment number looks good by comparison because it mixes two factors, jobs and the size of the labor

force, and neglects the fact that the labor force has contracted sharply. "A better gauge," he argues, "is the number of actual jobs." Three hundred thousand more jobs were lost in the 1991–92 recession than the 1982 recession, June 1981–January 1983. A larger portion of the jobs lost this time involved higher-wage white collar workers, with ramifications throughout the economy. People working or seeking employment has declined by 1.2 million people in the first 19 months of this recession as opposed to 125,000 in the first 19 months of the 1982 recession. These features are not revealed by the unemployment rate.

Manufacturing output is another statistical indicator of economic health. According to this statistic, using constant output values, manufacturing has remained near 22 percent of America's gross national product since World War II. But, McMillion notes: "Even Commerce Department officials who assigned these values admitted—in the Survey of Current Business last year—that "only a substantial research effort over many years holds any promise of overcoming . . . formidable statistical problems with these figures." The rapid pace of technological change makes it virtually impossible to measure "constant" output over time.

U.S. Competitiveness: A comparison of the gross domestic product per worker of the United States against that of other major industrial competitors shows the United States to be well ahead of such rivals as Germany and Japan. "The tally depends," McMillion observes, "on the value assigned to the dollar. . . . Most comparisons use theoretical—so called 'domestic purchasing power parity'— values that vastly over-value the dollar."

Clearly, the assumptions underlying such statistical economic indicators as unemployment, manufacturing output and U.S. competitiveness must be watched. What must also be watched are the limitations of such indicators, and what they omit which is relevant.

TV RATINGS: CAN YOU TOP THIS?

The life span of a television program is determined by the public's reaction to it, which is measured by TV ratings. These ratings, produced by the Nielson and Arbitron companies, estimate the audience in terms of the percentage of those sets in use which are turned to each channel, called a share, or in terms of the percentage of the total possible audience, sets on or off, called a rating. Shares and ratings are further broken down according to the sex and age of viewers so that advertisers can better focus their advertising campaigns. These numbers determine the buying and selling of billions of dollars of television air time. They mean life or death to television programs. The half-hour comedy *Good & Evil*, which had promising ingredients in terms of writing, acting and production talent, had a short life after its premiere in the fall of 1991 because of low initial ratings. In March 1992 NBC announced that they were dropping two successful shows, *Matlock* and *In the Heat of the Night*, because the demographic numbers favored older viewers while the network wished to build around a more youthful audience. (These programs were picked up by ABC and CBS, respectively, and continued through the spring of 1995.)

Since 1986 the data which underlie the ratings have been collected by a device called a people-meter. The remote control part of a people-meter rests on top of the television set. When the set is turned on, the meter prompts viewers to enter their identification number. Information is provided on what channels are being beamed into the household and who is watching them. Nielsen puts its people-meter into 4000 households selected at random—that is, without bias—from the approximately 93 million homes in America with television.

The people-meter data gathering system produced lower ratings for the networks than had been expected and a serious question arose as to whether this was because of the increased or decreased accuracy of this system over the method it replaced. The networks commissioned a study of the Nielsen methodology and two years later this Committee on Nationwide Television Audience Measurement (CONTAM) issued a nine-volume report that was highly critical of the Nielsen system. The report found evidence of button fatigue—that over time people did not push the buttons that would insure data accuracy as they did in the beginning. CONTAM was highly critical of Nielsen's sampling procedures for obtaining the 4,000 households that make up their sample; random sampling was envisioned in the methodology, but the actual sampling deviated significantly from this requirement. From this came ratings which were highly suspect. David Poltrack, senior vice president of research at CBS, observed that: "The whole business is crazy. I don't think there's an advertising agency in the United States that could get up in front of its clients and justify the way business is done right now. It's being bought on narrow based demographics, demographic targets which are not representative of product consumption in the United States."[6]

On the other hand, some advertisers view the accuracy of ratings numbers as being of little concern. Alice Sylvester of J. Walter Thompson advertising agency commented that: "The Nielsen ratings right now represent an agreement between buyers and sellers to use them as the currency of negotiation. The whole system comes down on supply and demand, and generally not what the individual ratings points are, so much like the commodities market, the prices are based on how many people want to buy their commodity. But for what we're doing with the ratings and the way we plan and buy and execute, it works just fine."[7]

It is not just fine, however, for the actors, actresses, writers and production companies to see their creative work go under on numbers which are, at best, suspect. Nor is it comforting for the viewing public to have its options determined by such numbers.

WHAT IS THE UNEMPLOYMENT RATE?

The "official" unemployment rate is what the Labor Department declares it to be at the beginning of each month. For a certain month the "official" rate is cited as 6.5 percent, let us say; what does this mean? It means that 6.5 percent is the figure determined by means of the official rules which define the calculation of the rate from the data obtained. The 6.5 percent figure is intended to provide a picture of unemployment in the country. The quality of the picture obtained depends on the quality of the "official" rules—the camera, the quality of the data obtained—the film, and the quality of the photographer—those who make up the "official" rules and carry them out.

Numerous critics of the picture of unemployment provided by the unemployment rate have argued that the picture is seriously out of focus because it does not record those who have become discouraged and have stopped looking for work, who are on the edge of unemployment, such as those who are temporary or freelance workers seeking full time work, who are working at jobs beneath their edu-

cation or skill level, who are older men and women pushed into lower paying jobs.

MODIFICATION OF THE CAMERA AND FILM. In November 1993 the Government publicly acknowledged that it had been substantially underestimating unemployment for at least a decade, particularly among women and discouraged workers. To obtain a more accurate picture of unemployment, revised camera and film, including the collection of additional data, revision of the questionnaire employed, changes in definitions, and the computerization of the entire process, beginning with employment of laptop computers by interviewers going into households, were put into operation in the beginning of 1994.

In its redesigned questionnaire, the Labor Department dropped what had come to be called its "most notoriously complex" question: "Did you have a job or business from which you were temporarily absent or on layoff last week?" It was replaced by simpler, more straightforward questions. The new questionnaire employs a series of yes or no questions to determine whether the interviewee is part of the work force. The interviewer is not asked to use judgment in shaping the questions. For further discussion of the importance of wording of questions, see *Watch the Questions, Among Other Things,* Chapter 8.

WHAT IS THE BUDGET DEFICIT?

When we think in terms of our personal finances, the answer seems straightforward enough. Add up income and expenses, and income minus expenses is a deficit if this difference is negative; it is a surplus if this difference is positive.

The Government's situation, however, is not so straightforward. "Discussion arises," as Charles Zwick and Peter Lewis point out,[8] "because the budget is not simply a way to divide up revenues but also a way to stimulate or cool down the economy by influencing purchasing power, growth, inflation, interest and exchange rates." Zwick and Lewis point to six questions whose answers will influence the size of the deficit.

1. Should interest payments be included in the deficit?

2. Is a deficit caused by a shortfall in revenues in a recession a true deficit?

3. Should capital expenditures be included in the same budget as ordinary expenditures?

4. Should the savings and loan clean-up be included in the budget?

5. Is the deficit reduced by inflation?

6. Should Social Security programs remain in a unified budget?

There are 64 possible budget deficits, depending on how these questions are answered. For fiscal 1993 the federal budget deficit could run as high as about $470 billion for an affirmative answer to all six of the preceding questions, or be a surplus around $150 billion for a negative answer to all six of the preceding questions.

8 ▶ COMING UP WITH THE NUMBERS

MORE EASILY SAID THAN DONE

It is often clear to the mind's eye what data are required. But the mind's eye does not have to contend with reality, which makes the task of obtaining the data more easily contemplated than accomplished.

CAN YOU TRUST POLLS?

The short answer, of course, is that it depends. The long answer brings us to what it depends on and what to watch for and beware of.

One of the most interesting and instructive cases involved what may be termed the polling Goliath and the three Davids. The polling Goliath that emerged in the period between the two world wars was *The Literary Digest*, a weekly publication similar in many respects to today's *U.S. News & World Report* and *Time*. In early 1920 the *Digest* mailed over eleven million ballots to obtain a sense of the public's view on possible presidential candidates. In 1924 it conducted its first presidential poll, sending out some sixteen and a half million ballots to people in all 48 states. For the 1928 and 1932 elections even more ballots were sent out. The mailings were massive, which by itself convinced many about the reliability of the poll. The hoopla accompanying the polls was deafening, and their overall success in predicting the presidential winners with a small margin of error was stunning. Some doubts were, however, raised. The mailing list for the ballots came largely from telephone directories and automobile registrations and some critics suggested that there was a class bias in the list which favored Republican candidates. This bias was pointed to as explaining why the predicted Republican vote was considerably higher than it turned out to be in a number of states. In 1932 the *Digest* poll predicted Roosevelt would win the popular vote with a margin of 59.85% and carry 41 states with 474 electoral votes. Roosevelt received 59.14% of the popular vote and carried 42 states with 472 electoral votes. It's hard to argue with success.

The 1936 presidential election was on the horizon with Franklin D. Roosevelt, seeking reelection, facing the Republican nominee Alfred E. Landon and minor party candidates. Riding high, *The Literary Digest* geared up. It mailed some ten million ballots to prospective voters and eagerly awaited the returns.

Against this background it seemed foolhardy in the extreme for George H. Gallup to take on Goliath. Gallup had founded the *American Institute of Public Opinion* in 1935 and initiated a weekly column called *America Speaks*. To attract newspaper subscriptions, he offered a money back guarantee that his prediction of the presidential

winner would be more accurate than that of *The Literary Digest*. Gallup predicted a Roosevelt victory with about 54% of the popular vote. In response to its poll, the *Digest* received 2,376,523 responses, with 1,293,669 for Landon, 972,897 for Roosevelt (42.9% of the two-party vote) and the remainder for third party candidates. The *Digest* had Landon carrying 32 states with 370 electoral votes and Roosevelt carrying 16 states with 161 electoral votes.

The actual vote gave Roosevelt a landslide victory with 62.5% of the two-party popular vote. He carried 46 instead of 16 states and received 523 instead of 161 electoral votes. Already in financial trouble, *The Literary Digest* suffered a devastating blow to its credibility and folded the following year.

Independently of Gallup, two other researchers, Elmo Roper and Archibald Crossley, using similar methods, had also predicted a Roosevelt landslide. The three Davids, particularly Gallup, and their more accurate polling methods, stood tall over the fallen Goliath humbled by reality.

LESSONS. The 1936 Roosevelt landslide carried with it a number of lessons for the practice of polling in general and election polling in particular.

SAMPLE SIZE. The accuracy of a poll is not determined by enormous sample sizes, impressive as they might appear to be. The major point is that the sample must be "properly" chosen. Pre-election poll samples these days tend to be between one and two thousand in size.

TARGET POPULATION VERSUS THE SAMPLED POPULATION. It is most important that the population actually being sampled be the one about which we seek to draw inferences. The failure of *The Literary Digest's* 1936 poll was in part due to the sampled population being significantly different from the target population of prospective voters. It was generally appreciated that samples drawn largely from telephone directories and automobile registrations favored wealthy Americans who tended to vote Republican, but the extent to which this might be the case was not generally appreciated. After all, *The Digest's* spectacularly successful 1932 poll drew from the same sort of lists.

What went wrong? In 1932 voters of all economic strata tended to vote against President Hoover and the Republicans, holding them responsible for the Great Depression and its aftermath. Economic class differences were obscured, and *The Digest* got lucky in its poll. In 1936 the upper economic *strata*, disturbed by the direction of President Roosevelt's *New Deal*, were much more willing to return to the Republicans, as reflected by *The Digest's* 1936 poll. Spectacular success without an understanding of its origins bred spectacular arrogance followed by spectacular failure.

NONRESPONSE ERROR. Of 10 million odd ballots sent out by *The Digest*, 2,376,523 returns were received for a 76% nonresponse rate. What were the nonrespondents thinking? It is hazardous to predict one way or another. Those who did respond wanted to make sure that their opinion was counted. Those who did not respond did not care about the poll. In some locations, such as Allentown, Pennsylvania, the sampling list was not drawn from telephone and automobile registration lists, but from voter registration lists with no inherent Republican bias. The data show a clear response bias in that the proportion who favored Landon in the poll was much higher than the proportion who favored him in the election. Those voters who came from the lower economic *strata* tended not to respond to *The Digest's* poll when invited to do so and they strongly favored Roosevelt.

Other lessons were to come, but the laboratory of reality required more time to make clear that additional fine tuning was required. The Dewey versus Truman election of 1948 provided some of these lessons.

THOMAS E. DEWEY VERSUS HARRY S. TRUMAN. Vice President Harry S. Truman became President on the death of President Franklin D. Roosevelt on April 12, 1945. It was a difficult post-war period and when Truman was nominated by the Democratic Party as its presidential candidate in 1948, he faced an uphill fight against the Republican challenger Thomas E. Dewey, Governor of New York. The polls showed Dewey comfortably ahead and he adopted a strategy of caution and platitudes ("Our future lies before us."), seeking to avoid making commitments and enemies. The political establishment, with the exception of Truman, felt that Dewey had it in the bag. Truman vigorously counterattacked, delivering 300 odd "give'em hell" speeches on a whistle-stop tour of the country. Truman went to bed early election day evening and woke up to find himself President for another term. The pollsters and political pundits had missed the boat.

FURTHER LESSONS. STABILITY OF VOTER OPINION. By early September of 1948 the polls showed Dewey ahead of Truman by at least ten points. The pollsters believed that public opinion was pretty well set by this time and would change little in the time remaining before the election. At this point they closed down their polling operations. Reality, of course, proved them wrong.

THE UNDECIDED VOTE. How is the undecided vote to be handled? Gallup split the undecided vote in the same proportion as those who had expressed a preference for Dewey and for Truman, which strongly favored Dewey. Subsequent analysis showed that 14 percent of the voters made up their minds in the last two weeks of the campaign and that 74 percent of these went for Truman. It is clearly hazardous to decide for the undecided.

SAMPLING METHODS. The major objective of sampling is to choose a sample which, in some sense, provides a good reflection of the attitudes of the larger population from which it is taken. With one sampling method, called quota sampling, an interviewer is instructed to interview an assigned number, or quota, of individuals in groups defined by specified characteristics such as race, religion, ethnic background, age, occupation, and economic status. Which individuals are chosen to be interviewed is left to the discretion of the interviewer, as long as the specified quota is met. With random sampling, on the other hand, it is envisioned that the sample is chosen in such a way that there is no bias, deliberate or inadvertent, which favors certain samples being chosen over others. Random sampling is an equal opportunity procedure, whereas the way that a quota is satisfied in quota sampling is dependent on the judgment and convenience of the sample taker. If it is desired that the sample be chosen in a way that group characteristics such as race, ethnic background, and so on, are reflected in the sample, but that the sampling within these groups be unbiased, then one may employ stratified random sampling. This method envisions the population from which the sample is to be drawn as consisting of non-overlapping components (such as the states of the United States, for example), called *strata*, where a part of the total sample is chosen at random from each *stratum*.

The quota sampling method employed by the three Davids in the 1936 election was a vast improvement over the mail poll conducted by *The Literary Digest* in that it yielded results more representative of the population at large and gave much better control over nonre-

sponse error. Gallup, for example, was much closer to the mark than *The Digest* in 1936, but his predictions were wrong on six states and he predicted a Roosevelt victory with about 54% of the popular vote, seven percentage points off the mark. It might not matter when the outcome is strongly one-sided, but in a close election it could make a significant difference. The Dewey versus Truman contest provided a decisive test case. It was a close election, no doubt. In electoral vote terms, Dewey would have won had he carried California, Illinois, and Ohio, each of which he lost by less than 1%. The size of Gallup's and Crossley's error in predicting Dewey's victory in 1948 (5.4 and 4.7 percentage points, respectively) was smaller than the error in predicting Roosevelt's victory in 1936 (7 percentage points for each). But in terms of the bottom line, wrong is wrong.

Polling's failure in 1948 brought matters to a head concerning the sampling method to be used. Random sampling was the technique favored by many critics, but it was more complicated and much more expensive to implement than quota sampling. Gallup had serious doubts that the increased accuracy achieved would be worth the expense, but after the 1948 election he and other pollsters switched to random sampling for pre-election polls.

THE FINAL GAP

"There's many a slip between the cup and the lip," the old saying goes, and this is especially the case with the gap between the last poll taken and the election itself. A poll may be likened to a picture taken by a camera at a point in time. Looking at a sequence of polls is analogous to looking at a sequence of still-photos, a videotape, if you will. While this might be strongly indicative of what will happen, it is not the same as what does happen. A reaffirmation of this point was provided by the British parliamentary election held on April 9, 1992. The polls consistently showed Labor ahead, but the Conservative Party won decisively. The British electorate wanted to send the Tories a message in the polls, but in the final moment of decision making, they felt they were better off with them than without them.

WATCH THE QUESTIONS, AMONG OTHER THINGS

WORDING. The wording of poll questions is, with the best of intentions, a delicate matter. In an experiment conducted by Elmo Roper as far back as 1940, for example, of those asked if the U.S. should forbid public speeches against democracy, 46% replied "no." Of those asked if the U.S. should allow public speeches against democracy, 25% said "yes." Support for free speech was much greater when the term "forbid" was used rather than "allow." "Do you think President Nixon should be impeached and compelled to leave the Presidency, or not?" In a Gallup poll held in July 1974, 24% of the respondents said "yes." When the question was posed as, do you "think there is enough evidence of possible wrong doing in the case of President Nixon to bring him to trial before the Senate, or not?", 51% said "yes." This was in response to a Gallup poll held at the same time. A case of fickle respondents? No; the second wording makes clear that Nixon would be brought to trial, which was not as clear to respondents from the first wording.

If you want the right answers, ask the right questions in the right way.

LOADED QUESTIONS. And then there are loaded questions worded to elicit a desired response for political purposes. Ross Perot's mail poll on National Referendum—Government Reform which appeared in the March 20–26, 1993 issue of TV Guide contains questions of this type. Question 13, for example, reads: "Should laws be passed to eliminate all possibilities of special interests giving huge sums of money to candidates?" The term "special interests" as it is used carries with it the ominous suggestion of special interests taking over the country. When stated in this form in a Time/CNN poll, it received an approval rating of 80 percent. The Time/CNN poll also put the question as follows: "Should laws be passed to prohibit interest groups from contributing to campaigns, or do groups have a right to contribute to the candidates they support?" Forty percent of the Time/CNN respondents stated they would prohibit interest groups from contributing, while 55 percent stated that they had a right to contribute.

OVERLOADED QUESTIONS. A study released by the Education Department in September of 1993 concluded that half of the adults in the United States cannot read or handle arithmetic. This is certainly an alarming figure, but is it accurate? That, to a large extent, depends on the success of the test makers in developing questions free of cultural bias, verbal ambiguity, and distracting irrelevancies. The following arithmetic question, for example, was posed: "The price of one ticket and bus for 'Sleuth' costs how much less than the price of one ticket and bus for 'On the Town'? A charter bus will leave from the bus stop (near the Conference Center) at 4 p.m., giving you plenty of time for dinner in New York. Return trip will start from West 45th Street directly following the plays. Both theaters are on West 45th Street. Allow 1 1/2 hours for the return trip. Time: 4 p.m., Saturday, November 20. Price: On the Town, Ticket and bus: $11.00; Sleuth: Ticket and bus: $8.50. Limit: Two tickets per person." The question itself raises the question of to what extent it is intended to test arithmetic and to what extent it is intended to test one's ability to successfully negotiate a verbal maze, particularly when many of those participating in the test are foreign born adults whose first language is not English. With tests of this sort, the underlying question of how well the questions achieve their intended objective must be given most careful consideration before meaningful conclusions can be drawn about the population being studied.

UNDERLOADED QUESTIONS. Can an unloaded question be loaded by virtue of being unloaded? There are those who would say yes in connection with a proposition to be put before California voters in November 1996. The proposition reads: "The state will not use race, sex, color, ethnicity or national origin as a criterion for either discriminating against, or granting preferential treatment to, any individual or group in the operation of the state's system of public employment, education, or public contracting." The proposition does not mention affirmative action, a term which evokes a strong emotional response, but if passed it would end affirmative action programs in California.

When put as stated to a representative sample of 800 California voters in a Harris poll conducted in May 1995, they favored it by 78 to 16 percent (6% undecided). When it was pointed out that the proposition would outlaw all affirmative action programs for women and minority groups, the pros of the sample went from 78 to 31 percent and the cons from 16 to 55 percent (14% undecided).

DIRECT QUESTIONS. Asking a direct question may provoke a misleading response. Do you intend to vote in the forthcoming election? Many people would answer yes rather than run the risk of being thought an irresponsible citizen.

QUESTIONS INVOLVING TECHNICAL TERMS. Questions involving terms which might be understood one way in everyday usage, but have a specific technical meaning, must be handled carefully. Robbery, involving confrontation between victim and offender, for example, is technically different from burglary, which does not involve personal confrontation. Such a distinction must be made clear in any question involving such terms.

ORDERING OF QUESTIONS. Not only may the wording of questions affect the response, but so may the order in which they are asked. A January 1984 prepresidential *New York Times*/CBS poll found that voters preferred incumbent President Ronald Reagan to Democratic challenger Walter Mondale by 16 percentage points; the question was posed at the beginning of the interview, which favored the better known Reagan. Gallup and *Washington Post*/ABC News polls taken around the same time posed the question near the end of the interview after questions about Reagan's policies had been asked; this helped Mondale because he had less of a record to defend. These polls had Reagan and Mondale about even.

RESPONSE OPTIONS. The two-category response option of the form Yes or No, employed in the aforenoted Perot poll, for example, is much more restrictive than the four category response option Yes, No, Not sure, Not enough information. The number of response options available may profoundly influence the response given, including the possibility of nonresponse. It is interesting to compare voter reactions in two July 1992 opinion polls on President George Bush and Democratic Party candidate Bill Clinton. The responses for the two polls are shown in the following tables.

	Bush	**Clinton**
Favorable	40%	63%
Unfavorable	53%	25%
Don't know	7%	12%

Table 8.1

	Bush	Clinton
Favorable	27%	36%
Unfavorable	49%	24%
Undecided	22%	31%
Haven't heard enough	2%	9%
No answer	1%	1%

Table 8.2

THE EFFECT OF BEING POLLED. Being polled itself may put people on guard if they feel that they might lose benefits or be penalized in some way if they give the "wrong" answers. In the 1930s, for example, many people on welfare were afraid they would be thrown off the welfare rolls if they gave an undesirable response. The election of Violeta Chamorro as President of Nicaragua in early 1990 was contrary to poll predictions which had Daniel Ortega with a substantial lead. One reason that the polls were so inaccurate was that the intimidation factor was not accurately taken into account. Nicaragua had been under authoritarian rule and in a state of civil war for a number of years, and many voters were not about to freely express their political preferences to pollsters. In such situations it is especially important to go all out to win the trust of respondents.

INTERVIEWER INDUCED BIAS. Conducting a "successful" interview, whether in-person or by telephone, is not a simple matter. People respond to interviewers as well as questions and the interviewer must strike a balance in being personable, respectful and considerate of the person being interviewed, and professional. The type of person who sets your teeth on edge by the way he says "Good Morning" is not likely to be a successful interviewer.

Needless to say, the more personal the questions, the more difficult will it be to obtain reliable responses. Questions probing a person's sexual behavior immediately come to mind in this regard.

America's democratic political institutions provided a fertile soil for the emergence of the practice of asking people what they think in some sort of organized fashion. If people are expressing their preference on election day, the day they would be asked their opinions beforehand could not be far behind. The first published presidential poll was published in *The Harrisburg Pennsylvanian* on July 24, 1824. Andrew Jackson won with 335 votes to 169 for John Quincy Adams. (Jackson subsequently carried the popular vote by a wide margin, but not the electoral vote. The election went to the House of Representatives which elected Adams President.) The practice of polling was thus launched. It was to undergo many extensions and refinements.

Today the practice of conducting polls, or surveys, as they are also called, is widespread with an enormous range. Political preference polls are perhaps the best known, but the majority of surveys that are conducted are concerned with specific social, commercial, or administrative concerns that do not occupy the public's spotlight with the same intensity. Thus, for example, government agencies conduct surveys to obtain information about employment, who uses food stamps, and who uses the national parks. TV networks employ surveys to obtain information about the popularity of their programming; businesses employ surveys to determine consumer preferences and reactions to their products; and interest groups conduct surveys to obtain a sense of the public's reaction to such matters as the state of education, the economy, crime, race relations, health costs, environmental issues, gun control, foreign policy issues, and the war against drugs.

The problems inherent in carrying out a poll, or survey, to obtain reliable data and project them on a larger population are clearly considerable, which should give us pause to closely watch the trouble spots and lead us to be cautious in interpreting the results obtained.

WERE THE NUMBERS "FAIRLY" DRAWN?

Random sampling, which envisions a sample being drawn without bias, deliberate or inadvertent, which favors certain samples being drawn over others—a level playing field, if you will—is for many activities the preferred sampling method. In a wide variety of situations it has come to be the definition of numbers "fairly drawn." While it is easily contemplated as an ideal, achieving random sampling in practice is not a simple matter. Consider the following situation, for example.

To raise money for Huxley College's scholarship fund, the student Social Science Society sold tickets at the College's homecoming affair. The prize was a 32-inch television set. The tickets were placed in a bowl and mixed. About midway through the festivities, Huxley's President, blindfolded, reached into the bowl and drew the winning ticket. What everyone expected and assumed was that the drawing was fair in the sense that there was no bias in the drawing which favored some tickets being drawn over others—that is, that it was drawn at random. On the face of it the procedure seems reasonable enough for the task at hand, but how close does it come to satisfying the requirements of a random drawing? The problem is with the physical stirring of the tickets to achieve a "thorough mix." Obtaining a "thorough mix" becomes more and more difficult to achieve as the number of tickets increases, and it is not clear whether early, middle, or late ticket buyers might be favored and to what extent. Still, for the task at hand the degree of randomness achieved might be random enough, except possibly for those individuals who are willing to go to war over a TV set or what they consider the "principle of the matter."

GOING TO WAR. When it comes to going to war based on the outcome of a random drawing, the stakes are raised considerably. The European phase of the Second World War and Japan's aggressiveness in Asia greatly alarmed the United States, and on September 6, 1940 Congress passed a conscription law, America's first peacetime draft. To implement the draft each eligible man in a Selective Service District was assigned a number which was put into a capsule. The capsules were put into a bowl, stirred, and then drawn one at a time. The order of the drawing determined the order of the draftees. The properties of the resulting sequence prompted questions about the randomness of the drawing, which gets back to the thoroughness of the mixing.

In 1969 the administration of the draft in the United States to determine the order in which men born in 1950 would be drafted was changed to a lottery system. Three hundred sixty six capsules were prepared (for a leap year), each containing a birth date. Each month's capsules were put into a separate box. The boxes were emptied into a drum, first those for January, followed by those for February, and so on for the subsequent months. The drum was rotated a few times, the capsules were poured into a bowl, and on December 1, 1969 the drawing was made. Those with birthdays on the capsules drawn first would be drafted first, and so on. If your birthday fell among those drawn last, there was a good chance that you would not be drafted at all. The results of the drawing are given in Table 8.3, from which we see that the earlier months, January through June, got the larger share of the last-to-be-drafted numbers and the later months, July through December, got the larger share of the first-to-be drafted numbers.

Month	1–122 (First-Drafted)	123–244 (Middle)	245–366 (Last Drafted)
January	9	12	10
February	7	12	10
March	5	10	16
April	8	8	14
May	9	7	15
June	11	7	12
July	12	7	12
August	13	7	11
September	10	15	5
October	9	15	7
November	12	12	6
December	17	10	4

Table 8.3

The results suggest the possibility that the earlier months' capsules were concentrated at the bottom of the bowl, while those of the later months were concentrated at the top and were more accessible for picking. Statistical hypothesis tests of randomness did not support the hypothesis that a random drawing had been carried out.

This period was one of great turbulence in American history and slogans such as "Draft Beer, not Students" and "Hell No, We Won't Go," were a prominent part of the scene. Perceptions of an indefensible draft for an indefensible war added fuel to a raging fire. In response to criticism, the Selective Service modified its number selection mechanism for the draft lottery conducted in 1970.

What these examples serve to make clear, however, is the difficulty of achieving a random selection in practice.

RANDOM NUMBER GENERATION. The idea behind random number generation is that the digits from 0 to 9 are selected by a process such that each digit selected is independent of any other digit selected and all digits have the same likelihood or chance of being selected. The process, usually based on a computer program, is set in motion and thousands of digits are generated and recorded in the order in which they are generated.

The study of many complex phenomena requires the generation of large streams of random numbers. It came as quite a shock when three scientists showed that five of the most often used computer programs for generating random numbers induced errors in the study of behavior of atoms in a magnetic crystal because the numbers produced were not random, despite the fact that they passed several statistical tests for randomness.[1] The deviations from randomness were subtle and, although the pseudorandom numbers produced were satisfactory for many purposes, they were not satisfactory for the problem at hand. Is it possible that no machine based system can produce truly random numbers? John von Neumann, regarded as the father of the modern computer, thought that the answer is yes. In an observation made in 1951, von Neumann expressed the view that anyone who believed a computer could produce truly random numbers was living in a state of sin. It may be that the best we can hope to do is produce pseudorandom numbers which are satisfactory for the purpose at hand, and that the truly random number is a mathematical ideal which cannot be attained. The question that arises in an applied situation then is, how random is random enough?

SEXUALITY BY THE NUMBERS, OR NOT?

In 1987 Shere Hite published *Women and Love: A Cultural Revolution*, her third book on human sexuality. In her first two books, *The Hite Report* (1976) and *The Hite Report on Male Sexuality* (1981), Hite restricted herself to telling what she had learned from women and men who replied to the extensive questionnaires concerning their sexual problems and attitudes she had circulated. For her first book she circulated approximately 100,000 questionnaires, from which she received 3019 responses for a response rate of about 3%. Four different versions of the questionnaire were sent to women's organi-

zations who were asked to circulate them. A similar methodology was employed for her second book. Approximately 119,000 questionnaires were distributed with a response rate just under 6% being obtained. Shades of the *Literary Digest's* debacle come to mind, but Hite was not claiming that her sample was representative of women and men in general. Hers was a qualitative rather than statistical study. In statistical studies, the same questions must be asked of all prospective respondents with the same response options being available to them all. Uniformity of the underlying conditions and the choosing of a "representative" sample so that the results obtained could be projected onto the population at large are essential for a statistical study. Qualitative studies, on the other had, focus on the special qualities of each individual potential respondent. Capturing the diversity inherent in individuals takes priority over ensuring uniform underlying conditions. It is not a matter of one kind of study being superior to the other, but rather of which methodology is appropriate to the study being undertaken. Although her first two books raised much controversy, Hite was on safe methodological ground.

In her third book, Hite attempted to cross the bridge from the qualitative results she had obtained to statistical generalizations about sexual attitudes of women in America. The bridge collapsed. Her methodology was almost universally criticized. ABC News in conjunction with the *Washington Post* conducted a telephone poll from October 15–19, 1987 to see if they could duplicate her results. They could not; their results were sharply at variance with those projected by Hite.[2] To take two examples, she found 84% of women as not being satisfied emotionally with their relationships; ABC/WP found 7% of married women and single women in a relationship as not being emotionally satisfied. Hite found 78% of women feeling they are only occasionally treated as equals in their relationships; ABC/WP found 81% of married women and single women in a relationship feeling they are treated as equals most of the time. There were differences in the way questions were posed, but they were not startling. As to who is closer to the mark, ABC/WP clearly takes the Trustworthy Prize because of its sound statistical methodology.

FOR "REAL" ACCURACY, COUNT THEM ALL. RIGHT?

A sample provides us with a view of the population from which it is drawn, an excellent view if it is "properly" extracted. But would it not be preferable to look at the entire population? Here, too, the short answer is that it depends; the long answer involves consideration of what it depends on. What it depends on includes the nature and size of the population we seek to get a grip on, degree of accuracy required, feasibility, and cost.

To count the number of defectives coming off a production line, a randomly chosen batch is inspected, the number of defectives in the batch is determined, and an extrapolation is made to the production line as a whole. This procedure gives sufficient accuracy, is cost effective, and timely. If there are good indications that the number of defectives is out of control, the production manager wants to know about it as quickly as possible so that the production process can be adjusted. Sampling accommodates this constraint. An audit of a company doing a high volume of business is accomplished by examining a randomly chosen sample of transactions and extrapolating to the whole. Examining every transaction could take an inordinate amount of time and resources with less accurate results.

To count the number of fish in a lake it is hardly feasible to do a direct count by catching them all and keeping track of how many were pulled in. In this situation, and others like it, a technique called capture-mark-recapture is employed. A large number of fish are caught, tagged, and released back into the lake. A reasonable amount of time is allowed to elapse to allow the fish to disperse (a few days, perhaps), another batch of fish is caught, and note is taken of how many were previously tagged. If, for example, 1 percent of the second batch were previously caught, the number of fish caught in the first sample is taken as an estimate, called the maximum likelihood estimate, of 1 percent of the number of fish in the lake, which gives us an estimate of the lake's fish population. If 200 fish, let us say, were caught in the first batch, 200 would be our estimate of 1 percent of the fish population, which yields a maximum likelihood estimate of 20,000 for the fish population.[3]

In seeking to determine the population of the United States and its characteristics in the mandated decennial census, the approach used in the last census of 1990 was similar to that used in the first census taken two centuries ago in 1790, direct count. It is based on the view that people are in a particular location at a particular time and that it is basically a matter of having them fill out the required questionnaire and then tabulating the results. In theory straightforward, in practice not so. The homeless, migrant workers, and people on the move are difficult or impossible to locate; the questionnaire is unclear to many who are located and asked to fill it out; those who refuse to fill out or return the census questionnaire are not insignificant in number (in some large cities response rates fell under 50 percent in the 1990 census); mispunching errors create whole categories of nonexistent people; some population subgroups are overcounted, while others are undercounted, and on it goes. The 1990 census found 248.7 million people living in the United States. A post-census analysis estimated four millon to six million people as not having been counted.

The discrepancy is much more than of academic interest because the census is the basis for determining how $60 billion of Federal money is divided up and legislative districts are redrawn. Challenges were issued to adjust the census figure, but then Commerce Secretary

Robert Mosbacher declined to "abandon a 200-year old tradition of how we actually count people." His decision was challenged in court, which subsequently found that the Secretary had not acted arbitrarily or capriciously in refusing to adjust the census figure. In a legal, if not numerical, sense we have an "exact" count after all. Alas, direct enumeration does not necessarily yield the accurate count in practice that one might believe it capable of in theory.

INDEX NUMBERS IN THE NEWS: HOW RELIABLE ARE THEY?

An index number is a number constructed to measure the change in a quantity over time, or how one quantity compares with another. While widely associated with the study of fluctuations in business and economic activity, they are employed in other fields as well. Index numbers continually in the news include the IQ value—which compares a person's intelligence score with an average for his age, the Index of Leading Economic Indicators—developed to alert us to changes in the nation's economic activity, and the Consumer Price Index (C.P.I.)—one of the most generally accepted measures of fluctuation in personal purchasing power.

The behavior of these index numbers affects the lives of millions because of the far reaching decisions they trigger. Changes in the Consumer Price Index, for example, trigger cost-of-living adjustments in social security payments for more than 43 million people, cost-of-living adjustments for workers whose wages are tied to the C.P.I., as well as adjustments in pensions of retired military and federal civil service personnel. The term "index numbers" itself communicates a sense of precision, but how precise are they? What are their limitations?

To obtain some insight into the nature and limitations of index numbers, consider a collection of basic clothing items used by a "typical" Huxley College student, listed in Table 8.4 The problem is to construct an index number that compares the overall current price of these items with their overall price in 1990.

	Unit Price	
Item	**1990**	**Current Year**
Shirt/blouse	$ 15	$ 25
Socks (pair)	1	2
Jeans (pair)	23	38
Shoes (pair)	50	65
Sweater	16	25
Total:	105	155

Table 8.4

One way to proceed is to add up current year prices ($155), divide this value by the sum of 1990 prices ($105), and multiply by 100 to express the ratio in percentage terms. This yields what is called a simple aggregative index for the current year (compared against the base year 1990).

We have:

$$I = \frac{155}{105} \cdot 100 = 147.6$$

We can interpret this result in two ways. We can say that it cost 47.6% more in the current year than in 1990 to purchase the group of items listed in Table 8.4. Equivalently, the current year aggregative cost of these goods is 147.6% of their 1990 value.

This procedure is simple to use, but has two major drawbacks. First, the index may be unduly affected by items having large price fluctuations. Suppose, for example, that Table 8.4 had listed a sixth article of clothing, a jacket whose price had dropped from $120 to $90 over the time span. Then the index would have been

$$\frac{245}{225} \cdot 100 = 108.9,$$

which suggests that there was only a small increase in overall prices, even though five of the six items in the group increased in price. The sharp change in the aggregative index was due to this sixth item having a much higher price in the current year and 1990 than the other articles of clothing in the survey. Suppose, instead, that the sixth clothing item, underwear, diminished in price from $5 to $2, which represents a larger percentage drop in price (60% drop) than the previous change from $120 to $90 per jacket (25.0% drop), but which involves smaller absolute dollar values. For this case the aggregative index would be:

$$\frac{157}{110} \cdot 100 = 142.7$$

This value is much closer to the original aggregative index, which is due to the use of smaller dollar values for the underwear and despite the steeper percentage drop in the price of the underwear compared to that for the jacket.

Another problem is that the aggregative index may be changed by altering the units of the items being compared; the more drastic the unit change and the greater the number of commodities whose units are changed, the greater the potential change in the aggregative index. If in Table 8.4, for example, we had expressed the unit price of jeans in terms of every ten pairs, then the 1990 and current year prices per unit would have been listed as $230 and $380, respectively. Our aggregative index would then be

$$I = \frac{25 + 2 + 380 + 65 + 25}{15 + 1 + 230 + 50 + 16} \cdot 100$$

$$= \frac{497}{312} \cdot 100 = 159.3$$

suggesting an overall price increase of 59.3% over the period in question, as opposed to the more modest estimate of a 47.6% increase obtained before for the same articles of clothing.

We can get around the units problem by employing what are termed price relatives. For each commodity, we compute the ratio of its price for the current year to that of the base year 1990. The value of this ratio is fixed, irrespective of the unit used for the commodity. We can then take the average of these price relatives to obtain an overall sense of the price changes and multiply by 100 to express the result in percentage terms.

From the data given in Table 8.4 we obtain the price relative ratios summarized in Table 8.5

Clothing	Price Relative
Shirt/blouse	$\frac{25}{15} = 1.667$
Socks (pair)	$\frac{2}{1} = 2.000$
Jeans (pair)	$\frac{38}{23} = 1.652$
Shoes (pair)	$\frac{65}{50} = 1.300$
Sweater	$\frac{25}{16} = 1.563$
Total	$= 8.182$

Table 8.5

Accordingly, we obtain for the price relatives index

$$I = \frac{8.182}{5} \cdot 100 = 163.6.$$

We interpret this result as meaning that prices have increased 63.6% on average for the relevant goods over the course of the time period. Not surprisingly, this result differs from the simple aggregative index.

The value of an index depends on the method employed, the years chosen for comparison, the commodities chosen for the index, and the regions chosen as a source of data.

As before, if one or few commodities undergo significant price changes, this may significantly alter the value of the mean of the price relatives index. For instance, suppose that the price of a pair of shoelaces tripled from $0.50 to $1.50 between 1990 and the current year. If we include this accessory item in our price relatives computation, the index becomes

$$I = \frac{8.182 + 3.00}{6} \cdot 100 = 186.4,$$

which is significantly larger than 163.6, the price relatives index obtained without the shoelaces.

On the other hand, if we include the shoelaces, the revised simple aggregative index becomes

$$I = \frac{156.5}{105.5} \cdot 100 = 148.3$$

which differs little from 147.6, the simple aggregative index obtained prior to inclusion of the shoelaces.

Many would consider the change in the latter index to be the more realistic in this case. Even tripling the cost of a pair of shoelaces should have a minimal effect on the overall change in the price of clothing because its cost is still low relative to that of other articles.

The addition of this low cost item into the price relatives calculation had a disproportionate effect on the index. If we took into account the quantities used of these articles, we would get a more realistic picture of overall price changes.

If getting a grip on overall price change in this comparatively straightforward situation offers such difficulties, what can we expect when we magnify the scale to the level of the nation's economic life or overall consumer purchasing power? In three words: much greater difficulties.

The Consumer Price Index, introduced during the first world war, underwent a number of significant refinements in 1940, 1953, 1964, and 1978. A view held by a number of economists and statisticians is that in its current form it overstates inflation. The amount by which it overstates inflation is in dispute. In late 1994 the Congressional Budget Office stated that the C.P.I. exaggerated inflation by an amount between 0.2 and 0.8 percentage points. At a joint meeting of the House and Senate Budget Committees held in January 1995, Alan Greenspan, head of the Federal Reserve Board, testified that the C.P.I. now exaggerates annual inflation by an amount between 0.5 and 1.5 percentage points. This overstated payments for Social Security and other cost-of-living payments tied to the C.P.I. which if corrected could save the Federal Government $150 billion over five years, according to Greenspan. A number of budget cutters greeted his suggestion with glee and began thinking about ways to implement a reduction in cost-of-living payments triggered by the C.P.I. without committing political suicide by offending the large voting group whose benefits are tied to this index.

What basis in fact do such estimates have? As the Huxley College situation makes clear, index numbers are bound to exaggerate one way or another the changes they have been designed to estimate. For a proper perspective on index numbers, it is important to keep in mind that they are tools of our making to help us describe and predict the movement of highly complex phenomena. Sometimes they do the job well and sometimes they miss the mark by a wide margin. The more complex the phenomenon, the more difficult it is to realistically capture and predict its behavior in terms of index numbers. Their success as a tool depends on our success in capturing with them the essence of the phenomenon they are intended to describe, our success in obtaining realistic data required for their construction, and our ability to understand and appreciate the limitations of these statistics which we ourselves have designed. To blame index numbers when they are wide of the mark in their predictions is analogous to blaming a spoon for doing a poor job in cutting meat. When an index number no longer does a satisfactory job, does the answer lie in revising it so that it gives an adequate representation of the phenomenon at hand? Makes good sense, many would agree. The answer surely does not lie in making an artificial adjustment to satisfy political pressure groups guided by ideological fury. The time to refine the C.P.I. has clearly come and the Bureau of Labor Statistics has been working on a revision which is due to be announced in 1998.[4-5]

The Index of Leading Economic Indicators is currently the Government's major forecasting tool for predicting economic expansions and slowdowns. On a number of occasions, however, it has given false signals about the economic behavior of the nation and its usefulness as a forecasting tool has been called into question by many economists. The time for its revision has arrived, many would argue.

ANTIDOTES FOR SLIPPERY FIGURES AND BOGUS NUMBER MONGERS

NUMBERS, NUMBERS, EVERYWHERE

The development of a quantitative mindset in our culture and technology which accompanied the explosion in quantitative and statistical applications during the twentieth century has had a powerful impact on the social sciences, business, economics, management, the health sciences and on the way that everyday affairs are conducted in general.

As we have seen, this development has also had a downside consisting of abuses in the use of numbers, fakery, deceptions, and misunderstandings. The answer to this pollution trail does not lie in discarding quantitative methods, which would be equivalent to discarding the baby with the bathwater, but in becoming knowledgeable about potential danger signs of number pollution.

HOW BELIEVABLE ARE THESE NUMBERS? TEST QUESTIONS

Application of the following test questions for figures to an article or presentation which relies on numbers or has numerical implications should put us in a better position to detect misleading figures and number fraud.

1. Is a source given for the figures? If the answer is no, then questions 2, 3, 4, and 8 are not viable. "Harper's Index," a column of *Harper's Magazine*, states percentages and figures for which

sources are cited, a practice which deserves emulation. Lester Thurow's op-ed piece "Companies Merge; Families Break Up" (*The New York Times*, Sept. 3, 1995; p. E11), to take another example, states percentages for which sources are not cited. This practice benefits figure finaglers and detracts from the credibility of perceptive analyses, like Professor Thurow's, which deserve serious consideration.

2. Is the source unbiased and reliable? What experience have I had with the source in the past? Has it earned membership in the Slippery Statistics Society (SSS), a reputation for reliability, or is its reliability status unknown?

3. How were the figures obtained? Does the method for obtaining the figures have weaknesses which may compromise the figures themselves?

4. How current are the figures?

5. Are the figures relevant to the issue in question? Is there more to the issue than they suggest?

6. Are there other figures which support the conclusion reached?

7. Are there other figures which contradict the conclusion reached? How are the contradictions to be resolved?

8. What assumptions underlie the numbers and conclusions obtained from them? Are these assumptions realistic?

9. Do the figures admit contradictory interpretations? If so, how are these contradictions to be resolved?

10. Do the arguments presented have quantitative implications? Do these implications cast doubt on the arguments?

STRENGTHENING THE SHIELD AGAINST BOGUS NUMBERS

Application of the aforenoted test questions to an article or presentation which relies on numbers or has numerical implications sets

up a shield which, to some extent, can protect us from bogus numbers. We can strengthen that shield by means of the following measures.

CHALLENGE NUMBER DROPPING. With number dropping having reached a feverish level in debates, presentations, and articles, one filter to help stem the flow of bogus numbers and their mongers is to publicly challenge the veracity of figures presented, as the occasion arises. The burden of proof of a figure's veracity should always be understood to be the responsibility of the presentor, not the presentee.

If this practice were implemented on a large scale, presentors of figures, data, and statistics would be more concerned about their veracity and bogus number mongers would lose their credibility through exposure.

WRITE A LETTER OF PROTEST. When you encounter a news-paper or magazine article in which numbers are bandied about freely without justification or figures are cited without an appropriate context to make them meaningful, drop a short note to the editor; it need not be typed. This is an initiative all of us can take. One letter might not seem like much, but many letters from us all have a cumulative impact which is significant.

PROMOTE NUMBER ETHICS. One of the most successful ways of reaching people is through an appeal to ethical behavior, which should include what might be called number ethics. Number, data and statistics manipulation, exaggeration, distortion, creativity—call it what you will—is not a victimless practice. We are all, directly or indirectly, victims. Product safety personnel, for example, who participate in the suppression of data which show their product in an unfavorable safety light are collaborators in the destruction, misery and death that their product subsequently unleashes because of safety flaws, as are the company executives who participate in the suppression. The GM case discussed in *Two for the Price of One* (Chapter 3) comes to mind in this connection, as do the Halcion and Silicone Gel episodes discussed in *These Data May Give You Nightmares* and *Women's Nightmares* (Chapter 3). Government and political figures who manipulate data for a "higher good" or "cause" as they see it, or for personal political advantage, to take another example, contribute to the undermining of our political, economic and social institutions, set the stage for unrealistic decisions being taken, and open the door to the waste of an incalculable amount of

resources. The cases discussed in *Soviet Defense Outlays?* (Chapter 2), *Top of the Line Deception* (Chapter 3) and *Number Magic* (Chapter 5) come to mind in this regard. All of us are casualties of such number manipulation.

David Stockman's reaction to his own role in number manipulation, discussed in *Number Magic*, is insightful and a source of hope. His conscience speaks to us about his role in number manipulation with a sense of regret and surprise that those who he thought should have known better put no stop signs before him. Education about the toll that number manipulation takes and the promotion of a sense of number ethics in our land would hopefully stir more consciences to action before rather than after the deed is done.

CALL IT WHAT IT IS. One obstacle to coming to grips with number lies is in the euphemisms that have been adopted to conceal the harshness of the practice. Terms such as "reworking" the figures, "massaging" the numbers, carrying out a "management adjustment" and "creative accounting" are more innocuous sounding than falsifying numbers and figure fraud. The sugar coating is so thick that one is in danger of losing sight of what is in the pill, which, of course, is the idea. The bottom line is that figure fraud is figure fraud, no matter how much sweetener is applied, and it would make matters clearer for all concerned to get it out in the open by calling it what it is.

STRENGTHEN THE LEGAL ARSENAL. Buffeted by waves of polluted or inconclusive data, it is uplifting to encounter those with integrity who refuse to ride the wave and by doing so run the risk of incurring the wrath of powerful forces. We meet five such individuals in these pages: Dr. Margot O'Toole, who defied a Nobel Prize winner *(No Lofty Pinnacle)*; Dr. Daniel Tripodi, who stood up to a pharmaceutical giant *(Push Comes to Shove)*; Dr. Erdem Cantekin, who refused to go all the way with the team *(If You Don't Agree, Shut Up!)*; Dr. Frances Kelsey, who saved the country from the thalidomide disaster *(Running the Stop Signs)*, and Professor Marvin Goodman, who rocked the political establishment by his testimony on C.I.A. pressure to which he was subjected to slant upward estimates of Soviet defense outlays *(Soviet Defense Outlays?)*.

These and other such knights in shining ethical armor are entitled to our deepest respect and admiration. But if they are to be effective in the public arena, the legal arsenal must be strengthened. Consideration of three specific cases makes clear weaknesses in the legal arsenal's ability to protect the public interest.

THE G.M. CASE. As noted in *Two for the Price of One,* in February of 1993 a state court in Atlanta held General Motors liable for the death of a teenager in the crash of one of its GMC Sierra pickup trucks.

Ronald Elwell, a former GM safety engineer, provided heavily damaging testimony when he told the jury that GM had known for years that the truck's "sidesaddle" design of its fuel tank placement was defective. The jury viewed videotapes of GM's own crash tests which showed that when a pickup was struck on the side by another vehicle moving at 50 miles per hour, in almost every case the pickup's fuel tank broke open.

But what about the public interest in general? Pickups of this type are still on the road. Does one have to be killed or seriously injured before an action can be initiated, and then only in the name of the injured? Data suppression relevant to public safety should by itself be actionable and expose a company to stiff fines and product recall in the name of the public if the public's general interest is to be protected.

HALCION. As noted in *These Data May Give You Nightmares,* critics contend that the Upjohn Company, which makes the sleeping pill halcion, concealed data from the Food and Drug Administration that show that the pill is more dangerous than other sleeping medications and more likely to lead to memory loss, depression, paranoia, and hallucinations. The strength of the evidence, which was revealed as a result of a lawsuit, prompted the British, Dutch, and Belgian counterparts of the F.D.A. to ban halcion in their countries.

Those individuals who can establish in court that they suffered injury will receive damages. But, as in the GM case, what about the public interest if the drug received marketing approval under false pretenses in the form of data suppression or manipulation?

THE NBC—GM TANGO. As noted in *Two for the Price of One*, after losing the court case, GM announced its intention to sue NBC for rigging the evidence on its Dateline NBC program of 17 November 1992. This program showed a simulated crash involving the type of truck that was at the heart of the suit in progress against GM. The fuel tank of one of the trucks exploded, but the test was rigged in that ignition devices had been taped to the trucks to ignite a fire if the simulated crash did not produce the desired result. Viewers, however, were not informed of this adjustment.

NBC apologized profusely on the air to GM, eating its share of humble pie and crow. GM accepted the apology, dropped its suit, and resumed its advertising on NBC News; the tango had been completed. But what about the public's interest. NBC clearly crossed the boundary between information and manipulation and the public was a

victim. Laws that would recover damages for the public would help deter so-called news organizations from victimizing the public by means of the most blatant form of data manipulation.

DEBUNK MATH MYTHS. A major math myth, which has led to much misunderstanding, is that arguments which employ figures must be more precise, and thus more credible, than those which do not. The mathematical precision of the tool mistakenly suggests to many that arguments which employ the tool must, by virtue of the fact, carry a certain precision. This is a form of suggested respectability by association. If Herman is seen in the company of the rich, wealthy, and powerful, doesn't this suggest that he too is rich, wealthy, and powerful? If a point of view is seen in the company of numbers, which are in some sense precise, doesn't this suggest that it too must be precise and thus credible? The suggestion in both cases is a powerful one which has often been exploited to gain advantage. It is, of course, fallacious. With arguments employing numbers, we must look further to determine whether they are reliable, relevant and properly employed.

POLLS. Polls have become so numerous that Russell Baker's suggestion that "what this country really needs is a poll-ban treaty"[1] begins to sound more and more appealing, especially as election time approaches. Even if polls strongly supported this suggestion, they are likely to become more rather than less numerous. The problem centers on what we can do to help us distinguish polls that have been properly carried out from their flawed cousins. Consider the following:

1. Focusing on poll results without an appropriate contest is misleading. News accounts of poll results should give information about the date the poll was taken, sample size, survey design, percentage of respondents among those contacted, response options, random sampling error and what it means, and the complete wording and context of any questions that are discussed. Out-of-context poll results are best viewed with questioning skepticism.

2. It is useful to check several polls against each other. Are the results consistent? It not, try to find out why.

3. Watch for trends. If a polling organization takes a number of polls in sequence, do the results point to a trend?

4. An increasingly popular, but seriously flawed, polling method is the telephone poll in which the public is invited to call one number to register approval of a candidate or position and another number to register disapproval. Among other things, telephone polls of this sort lend themselves to gross manipulation by pressure groups and give us no handle on the opinions of nonrespondents.

EDUCATION. As is generally appreciated, good ingredients are required for a successful drink, but a stirrer is needed to pull them together. The stirrer in this case is education.

The first step is to heighten awareness of number pollution among those on the front lines of news reporting and analysis—reporters, journalists, columnists, analysts, editors in the print and broadcast media. Better armed, our frontier guards would be able to adopt "they shall not pass," Marshal Henri Pétain's dictum for advancing German forces in the First World War, as their guiding principle for slippery numbers and numerical Talosian images.

Much the same might be said for the general public. The best defense against having numerical wool pulled over our eyes is education. Talks, discussion groups, courses, and readings would be helpful. There is no reason why courses centered on slippery numbers and Talosian numerical images cannot be developed and given as early as the secondary school level. Alas, this kind of education has not been a concern of mathematics education organizations and the mathematics education community at large.

In addition to this work, I should like to call attention to the following books as useful allies for any number awareness and anti number-pollution campaign.

1. W. J. Adams with R. B. Adams, *Get a Firmer Grip on Your Math* (Dubuque: Kendall/Hunt, 1996)

2. S. K. Campbell, *Flaws and Fallacies in Statistical Thinking* (Englewood Cliffs, N.J.: Prentice-Hall, 1974)

3. R. Hooke with J. Liles, *How to Tell the Liars from the Statisticians* (New York: Marcel Dekker, 1983)

4. D. Huff with I. Geis, *How to Lie with Statistics* (New York: W. W. Norton, 1954)

5. A. J. Jaffe, H. F. Spirer, *Misused Statistics* (New York: Marcel Dekker, 1987)

6. J. A. Paulos, *Innumeracy* (New York: Hill and Wang, 1988)

7. J. A. Paulos, *A Mathematician Reads the Newspaper* (New York: Basic Books, 1995)

8. R. Reichard, *The Numbers Game: Uses and Abuses of Managerial Statistics* (New York: McGraw-Hill, 1972)

[4] is the classic in the field. Highly readable and enjoyable as well, it is still available in what must be at least its twelfth printing. Unlike its companions on this list, [1] provides food for thought questions and exercises to help one get a reinforced hold on slippery numbers, Talosian numerical images, and the mathematical modeling process in general.

PART TWO

GET A GRIP ON
MATH MODELS

MATHEMATICS FOR A VACATION TRIP

ANN'S MODEL

Members of the Adams family were recently engaged in planning their trips from home in Brooklyn, New York to the vacation town of Kennebunkport, Maine. Ann planned to make the trip in late July with a major stop at Putnam, Connecticut for the annual picnic held there. Ann's problem was to set up a mathematical representation or model of the situation which would enable her to predict the total time required for the journey.

The setting of any such problem presents numerous features and characteristics, many of which are irrelevant or unessential to the focus of the problem. In developing her model Ann had to sort this out and decide on which features were negligible. This required discretion and judgment, the most controversial aspect of the math model development process; one person's essential might be another's irrelevancy.

Ann examined a map and laid out a route. She made assumptions about the traffic flow to be expected along various points, speeds that would be possible, and the number of rest stops to be made and their duration. These considerations led her to a model consisting of two line segments joining points representing Brooklyn, Putnam, and Kennebunkport, the sum of whose lengths is 350 miles, and the problem of determining how long it would take an object moving at an average speed of 50 miles an hour to travel this distance. She envisioned an approximate 2 hour stay at the picnic in Putnam before continuing to Kennebunkport.

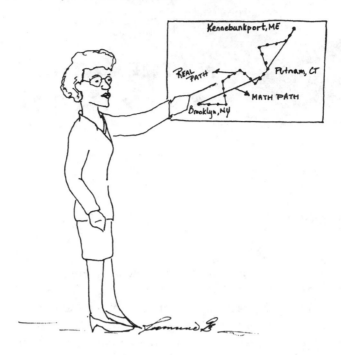

Ann's mathematical model is an idealized, abstract rendering of the real situation involving a trip from Brooklyn to Kennebunkport with a stop at Putnam. It is intended to capture the main features involved in taking such a trip and reflects these features as she sees them and the assumptions that she was led to make. It is possible that someone else planning such a trip would see things in another light and make very different assumptions.

By employing the mathematical operation division, we obtain the valid conclusion that an object moving along the idealized path of Ann's model at an average speed of 50 miles per hour would take $350/50 = 7$ hours to make the journey. If we add 2 hours for the picnic, we have a total of 9 hours for the travel time. This conclusion is a valid one with respect to Ann's model, valid in the sense that it is an inescapable consequence of the assumptions she was led to make. If we accept Ann's model as a starting point, then we must accept this conclusion as following from it in a deductive logical sense; it is inherent in the model. Mathematical methods, whether simple or the ultimate in technical sophistication, give valid conclusions from the mathematical models we set up. Valid conclusions obtained from a mathematical model are called theorems and the assumptions underlying the model are called postulates or axioms.

The interpretation of Ann's theorem is that if the trip were made under conditions realistically described by her postulates, it would take around 9 hours to complete the journey. What is the acid test of the accuracy of Ann's theorem—accuracy in the sense of how closely it describes the journey in the real world? Take the trip and see how long it takes. This is just what Ann did. It took her 9 hours and 10 minutes, which is close to the projected 9 hours for the journey. This established that her theorem is realistic in this case and, by reflection, was evidence in support of the realism of her model for the trip.

ANDY'S MODEL

Ann's son Andrius, called Andy by his friends, was planning his own trip to Kennebunkport to be taken in August. Andy did not intend to stop at Putnam and laid out a different model. Andy's assumptions about traffic flow, speeds that would be possible, and the duration of the one rest stop that he envisioned led him to a model consisting of a line 330 miles long joining points representing Brooklyn and Kennebunkport and the problem of determining how long it would

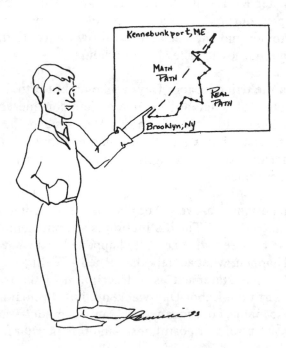

take an object moving at an average speed of 55 miles per hour to cover this distance.

By again employing the mathematical operation division, we obtain from Andy's model the valid conclusion that it would take 330/55 = 6 hours. This conclusion is valid with respect to Andy's model in the sense that it follows as an inescapable consequence of his model; it is inherent in Andy's model.

The interpretation of Andy's theorem is that if the trip were made under conditions realistically described by the postulates of his model, it would take around six hours to complete the journey. Andy took the trip in the middle of August and it took him 5 hours and 55 minutes, which is close to the projected time stated by his theorem. This established that his theorem is realistic in this case and, by reflection, was evidence in support of the realism of his model for the trip.

RASA'S TRIP: WHO'S RIGHT?

Andy's sister Rasa was planning to take a brief vacation trip to Kennebunkport during Labor Day weekend. Andy's model looked reasonable to her and she decided to follow the route it prescribed, expecting the journey to take around 6 hours.

Rasa took the trip on Labor Day weekend, but it took her 7 hours and 20 minutes. This actual trip time differs considerably from the projected 6 hour trip time of Andy's theorem so that something clearly went wrong; but what? "Your theorem stinks," Rasa shouted at her brother in a somewhat agitated manner; "it's wrong; it's not valid," she continued.

Rasa's experience proved Andy's theorem wrong in terms of reality, but not validity. The distinction is a fundamental one. In the confrontation between what actually happened and what the theorem says should happen, what actually happened—reality—wins. Andy's theorem is a false statement as a description of the travel time to Kennebunkport on a Labor Day weekend, but it remains a theorem. It is an inescapable consequence of Andy's model—more specifically, his assumptions or postulates—and this is what makes a theo-

rem a theorem. Mathematical methods, in this case the division of 330 by 55 yielding 6, did what it is capable of doing; it yielded a valid conclusion with respect to Andy's model. That valid conclusion may or may not be realistic. Validity is not the same as reality, but they are linked in this way: If the postulates of a model are realistic, so will be the valid conclusions obtained from it. If a valid conclusion is found to be false, this sends us a signal that some of the postulates of the model are unrealistic and require modification.

Andy's model, Rasa's experience showed, is not realistic for travel to Kennebunkport on a Labor Day weekend. In reexamining Andy's postulates we find that they do not realistically take into account unusually heavy traffic delays around the tollgates of the Whitestone Bridge characteristic of holiday weekends. Further examination of Rasa's actual trip shows that this is precisely where she had the difficulty.

We should keep in mind too that Andy did not design his model with a holiday weekend trip in mind. Major lesson: Look before you leap! That is, look at a mathematical model's assumptions before you use it.

Ann's and Andy's experiences showed that their theorems are realistic for their trips. By reflection, this is evidence in support of the realism of their models for the envisioned trips. This evidence does not, however, establish that their models are completely in accord with reality, as Rasa distressingly discovered when she employed her brother's model.

A mathematical model gives us a picture of a situation or phenomenon, but we cannot view the model as complete or the only possible picture. Other models are possible as well.

MATHEMATICS FOR BUSINESS AND ECONOMICS

THE BIRTH AND REBIRTH OF LINEAR PROGRAMMING

The seed from which the discipline called linear programming first germinated was planted in the late 1930s when the Leningrad Plywood Trust approached the Mathematics and Mechanics Department of Leningard University for help in solving a production scheduling problem of the following nature. The Plywood Trust had different machines for peeling logs for the manufacture of plywood. Various kinds of logs were handled and the productivity of each kind of machine (that is, the number of logs peeled per unit of time) depended on the wood being worked on. The problem was to determine how much work time each kind of machine should be assigned to each kind of log so that the number of peeled logs produced is maximized. A basic condition which had to be satisfied is that if logs of a given type of wood make up a specified percent of the input, then peeled logs of that type would also constitute that percent of the output.

The germination of this seed is due to Leonid Kantorovich, who saw that it together with a wide variety of economic planning problems can be formulated in terms of what are today called linear program models. These problems involved the optimum distribution of worktime of machines, minimization of scrap in manufacturing processes, best utilization of raw materials, optimum distribution of arable land, optimal fulfillment of a construction plan with given construction materials, and the minimal cost plan for shipping freight from given sources to given destinations.

167

A simple example of a linear programming problem or linear program involving two variables, x and y, let us say, is to find the largest value of P = 150x + 82y, where x and y are required to be non-negative and satisfy the linear constraint 2x + 3y ≤ 120. More generally, a linear program, which may involve hundreds and even thousands of variables, is a problem requiring that we find the largest or smallest value of a linear function where the variables must satisfy linear constraints[1].

In 1939 Kantorovich published a report[2] on his discoveries that included a method sufficient for solving all the linear program models he had formulated for the aforenoted problems. The chaos of the Second World War and the postwar intellectual climate in the Soviet Union were highly unfavorable for the development and implementation of Kantorovich's linear programming methods in the Soviet economic scene. Independently of the Soviet scene, linear programming methods were developed in the United States and Western Europe in the late 1940s, and the 1950s and 60s saw the development of a wide variety of linear program models for problems arising in such areas as economic planning, accounting, banking, finance, industrial engineering, and marketing. The thaw in the Soviet Union's intellectual climate which followed Joseph Stalin's death in 1953 saw the development and implementation of Kantorovich's ideas in the economic life of the U.S.S.R.

In 1975 Kantorovich was a co-recipient of the Nobel Prize in economics for his development of linear programming methods and their application in economics.

A TALE OF TWO LINEAR PROGRAMS

The Austin Company, a producer of high quality electronic home entertainment equipment, has decided to enter the digital tape player market by introducing two models, DT-1 and DT-2. Their problem is to determine the number of units of each model to be produced to maximize profit.

The Company's operations research department was asked to study the situation and make recommendations. The OR department began their analysis by collecting data. They divided the manufacturing process into three phases; construction, assembly, and finishing. The data collected and their analysis led them to make the following assumptions.

(a) In the construction phase each DT-1 unit requires 2 hours of labor and each DT-2 unit requires 3 hours of labor. At most 1,100 hours of construction time are available per week.

(b) In the assembly phase each DT-1 unit requires 5 hours of labor and each DT-2 unit requires 3 hours of labor. At most 1,400 hours of assembly time are available per week.

(c) In the finishing phase each DT-1 unit requires 4 hours of labor and each DT-2 unit requires 1 hour of labor. At most 756 hours of finishing time are available per week.

(d) After taking cost and revenue factors into consideration, the anticipated profit for each DT-1 unit is $150 and the anticipated profit for each DT-2 unit is $120. In order for these unit profit values to hold, the Company must produce at least 25 DT-1 and 40 DT-2 units per week.

(e) There is an unlimited market for the DT-1 and DT-2 models.

(f) All factors other than the ones considered in the analysis of the production of the DT-1 and DT-2 models are negligible.

These assumptions were formulated as a linear program model called LP-1[3]. A valid consequence of LP-1 is that to obtain the largest possible profit, $51,000 per week, the weekly production schedule should be set to manufacture 100 DT-1 and 300 DT-2 units.

Entry into the digital tape player market was a critical move for the Austin Company, which involved a considerable investment, and it decided not to take any chances. It hired the Marks Company, a consulting operations research firm, to independently study the digital tape player situation and make recommendations. The Marks OR group divided the manufacturing process into two phases: construction (which included assembly) and finishing. The data collected and their analysis led them to make the following assumptions:

(a) In the construction phase each DT-1 unit requires 8 hours of labor and each DT-2 unit requires 5 hours of labor. At most 2,210 hours of construction time are available per week.

(b) In the finishing phase each DT-1 unit requires 3 hours of labor and each DT-2 unit requires 2 hours of labor. At most 860 hours of finishing time are available per week.

(c) The anticipated profit for each DT-1 unit is $140 and the anticipated profit for each DT-2 unit is $150. In order for these unit profit values to hold the company must produce at least 50 DT-1 and 50 DT-2 units per week.

(d) There is an unlimited market for the DT-1 and DT-2 models.

(e) All factors other than the ones considered in the analysis of the production of the DT-1 and DT-2 models are negligible.

These assumptions were formulated as a linear program model called LP-2[4]. A valid consequence of LP-2 is that to obtain the largest possible profit, $60,250 per week, the weekly production schedule should be set to manufacture 50 DT-1 and 355 DT-2 units.

Now what? It's always easier when you have a choice of one; take it, or leave it. But a choice of two is another matter. Bottom-line Bob, chairman of the ten member board charged with making a decision on how to implement the Company's entry into the digital tape player market, argued that it's obvious what we should do. "Implementation of LP-2 brings us a weekly profit of $60,250, whereas implementation of LP-1 brings us a weekly profit of $51,000. Since we want the largest possible return, we should go with LP-2. You can't make it any simpler." The board voted nine to one to implement LP-2.

Alas, the $60,250 weekly profit was far from being realized after LP-2 was implemented, and two years later the Austin Company's venture into the digital tape player market had to be written off as a disaster.

Bottom-line Bob, who presided over this disaster was confused, upset, and out of a job. He went to Reflective Ramuné, the chair of the new board and the one person who had voted against implementation of LP-2, with some questions: "We were ultra-cautious and

obtained the additional services of the Marks OR group to make recommendations which, subsequently, had disastrous consequences for us; what went wrong? How is it that mathematics failed us? Why did you vote against implementation of LP-2?" "Well Bob, as I pointed out at the board meeting, I voted against implementation of LP-2 because I was not convinced that its promise of a $60,250 weekly profit was realistic. As you yourself pointed out, 'the promise of LP-2 is $9,250 better than that promised by LP-1,' but promises may not be realizable if they are founded on unrealistic assumptions. The conclusion reached from LP-2 was indeed tempting, and in fact proved too tempting for my colleagues on the board, but since it came from a linear program founded on assumptions which I viewed as unrealistic, I resisted temptation. We have no quarrel with mathematics; mathematics gave us a valid conclusion from LP-2, which is all that we can legitimately expect. Unfortunately that conclusion proved to be unrealistic."

IS MATHEMATICS PRECISE?

"Ramuné, I don't understand this. I always liked math in high school and college. Solving those equations, factoring those expressions, differentiating those functions, throwing the data into the computer and letting it do its thing, that was real fun. What I like most about math is its precision. You don't get ten sides to a story. You get one answer and that's that; no baloney.

"Bob, I think your math courses may have focused too much on technique and not enough on perspective. Technique can be fun to a point, but without a perspective on its place in the overall role of mathematics in applications, we see only a small tip of the mathematical iceberg. Mathematics is precise in the sense that it gives us valid conclusions based on the assumptions made, which is where technique—factoring, differentiating functions, and the like—plays its major role. Whether the assumptions made are realistic or not is another matter which technique can't help us with. The question of how to formulate these assumptions and reach a judgment on their realism may indeed yield ten sides to the story. I'm afraid that those who find mathematics attractive because of what they perceive to be its absolutist nature have misunderstood the meaning of mathematical precision."

FALLING SHORT: THE ADVERTISING MEDIA SELECTION PROBLEM

The advertising media selection problem is to chose from various advertising media a selection which is, in some sense, best. Potential advertising media include newspapers, magazines, radio, television, and direct mailing. In most variations of this problem, the objective is to maximize audience exposure. Budget is one of the constraints on the media selection to be made. Other restrictions on the kind of media mix selected might come from company policy, contract requirements, and limited availability of media.

In 1961 the firm of Batten, Barton, Durstine, and Osborn (BBD&O) came out with a linear program model for the media selection problem[5], which was followed by the publication of a number of articles on the application of linear programming methods to this problem. In the early 1960s, hopes ran high in the world of advertising for linear programming and, in a spasm of optimism, BBD&O placed full-page ads reading, "linear programming showed one BBD&O client how to get $1.67 worth of effective advertising for every dollar in his budget."

Initial optimism for linear programming turned to disappointment which in turn gave way to a calmer understanding of the potential and limitations of linear programming for media selection. This attitude transition is reflected in the examination, in a series of papers, of the realism of the assumptions which had served as a foundation for the initial linear programming approach to media selection. Philip Kotler[6] noted the following limitations.

1. Linear programming assumes that repeat exposures have the same effect.

2. It assumes constant media costs (no discounts).

3. It cannot handle the problem of audience duplication.

4. It says nothing about when ads should be scheduled.

Frank Bass and Ronald Lonsdale[7] explored the operational details of the application of linear programming to media selection and found linear programming models to be "crude devices to apply to the media selection problem . . . The linearity assumption itself," they concluded, "is the source of much of the difficulty. Justifying an assumption of linear response to advertising exposures on theoretical grounds would be difficult . . . Assumptions about the nature of response to advertising cause most difficulties in models of the type examined in this article." Later linear programming approaches to the media selection problem took into account the criticisms that had been voiced and sought to overcome them.

We should be careful to note that the criticisms voiced were not directed toward technical mathematical details, such as the conclusions reached or the mathematical techniques used. The solutions obtained were indeed valid with respect to the linear programs developed. If you accept a linear program model for a given media selection situation, then you must also accept the solution which inevitably follows from it. But must you necessarily accept a linear program model? If you accept the assumptions on which the linear program is based, the answer is yes. If you don't, the answer is no. The criticisms voiced were directed toward the assumptions that served as a foundation for the linear programming approach, and the authors of these criticism were saying: Be careful; for such and such reasons these linear programming assumptions are not realistic. Therefore you should be cautious about accepting a linear program model that has been developed to reflect them. Since it is founded on unrealistic assumptions, the conclusions which inevitably follow, while valid, may not be realistic.

The development of linear program models for the media selection problem illustrates an evolutionary factor which should be kept in mind for a correct perspective on the subject. The later linear programming approach to media selection was based on more realistic assumptions than the earlier approach, one which was made possible by critical examination of the earlier. A critical examination of a mathematical model's assumptions is positive in the sense that it paves the way for the development of refinements based on more realistic assumptions. Quite often, in fact, a mathematical model is initially proposed for a situation with the understanding that its assumptions are unrealistic. The value of such a model is that it serves as a prototype from which more realistic models can be developed.

At the present time there are a number of approaches to the advertising media selection problem. Dennis Gensch[8] reviews, compares, and critiques several of these approaches, including linear programming. His focus is on the realism of the assumptions which underlie the approaches he considers.

12 ▶ MATHEMATICS AND THE COMPUTER

HOW COULD IT BE WRONG? I USED A COMPUTER

"Ramunė, I still don't fully understand what went wrong. The company just spent millions to update its computer system. I had access to the latest and the best. Why didn't this save us from disaster." "Bob, Henry Clay's observation that 'statistics are no substitute for judgment' applies equally well to the computer. We cannot expect the computer to employ technological alchemy and convert unrealistic assumptions into golden truths. Keep in mind the

GIGO principle; if garbage in, then garbage out. Indiscriminate use of computer technology has made possible the generation of more and more nonsense more quickly than ever before by more people having less and less understanding of what they are doing."

THE COMPUTER'S RIGHT OF WAY

"If what you say is true, Ramuné, then what good is this super computer technology to us?" "For number crunching and delivering results quickly and efficiently, the computer is without equal, Bob. In this dimension it is the undisputed master of the field. The mathematical model building process and computers have developed a symbolic relationship in that computers have made it possible for us to solve previously unapproachable large scale problems that come out of mathematical models, while the accessibility of such problems to computer solution has made possible the use of such complex models. Unfortunately, none of this overrides the GIGO principle.

IF ONLY COMPUTERS COULD THINK

"Ramunė, what we need are computers that can think. Computers are based on a logic which allows for sharp unequivocal alternatives such as yes/no and true/false. Life is complex with many shades of grey in-betweens." "Work has been done in an area called fuzzy logic, Bob, which has been useful in addressing some of the in-betweens required of mechanical controls for an office building's elevator system, for example.[1] No matter what developments are forthcoming in this area, we will still not be relieved of doing our own thinking about what we put into our computers."

THE MATHEMATICS OF FREE TRADE

SOME SAY YES AND SOME SAY NO

"For once, we're asking you not to believe in fairy tales," the ad implored. "Six presidents of the United States, plus 41 governors—Democrats and Republicans alike—think that NAFTA (North American Free Trade Agreement) is good for America. NAFTA will increase employment in the U.S., NAFTA will encourage Mexico to adhere to strict environmental regulations, NAFTA will help stem illegal immigration into the United States."[1]

A contrary ad, "8 Fatal Flaws of NAFTA," argues that "NAFTA will eliminate tens of thousands of U.S. jobs and . . . will also assault the laws that protect our forests, wildlife, air and water, and keep our food safe to eat. Don't call it 'free trade'; it's corporate aid."[2]

Faced by such contradictory claims, how are the likes of us who are trying to distinguish fairy tales from substance to decide? My first question is, what can mathematical modeling and the figures tell us?

WHAT DO THE ECONOMIC MODELS SAY?

It depends on the assumptions. If the focus is on long term gains in economic efficiency where "full employment" and "perfect competition" are assumed, the model predicts small income gains around 0.1 percent for the United States and 2 to 3 percent gains for Mexico.

If one subscribes to a macroeconomic model for the United States economy as a whole which assumes that Mexico will run trade deficits with the United States, the valid conclusion is that the U.S.

stands to gain around 170,000 jobs over the 1990s. If one favors a macroeconomic model which assumes that Mexico will run trade surpluses with the United States, the valid conclusion is that the U.S. stands to lose around 420,000 jobs over the 1990s.

As to Investment/Employment, if one assumes that investment in Mexico does not reduce investment in the United States, it follows that no jobs are lost. If one assumes that investment in Mexico will reduce investment in the United States, it follows that as many as 290,000 jobs may be lost.

THE DEVIL'S IN THE ASSUMPTIONS

Which of these scenarios should we opt for, if any? The Devil's in the small print, the old wisdom has it, and in this case that translates to the basic underlying assumptions of the models. If we accept a model's assumptions, then we are stuck with its conclusions, like them or not. Must we accept a model's assumptions? If they are "sufficiently realistic," it makes good sense to do so; if not, then no. But what does "sufficiently realistic" mean? There is, alas, no sharp, unequivocal answer. It will, in the final analysis, be a judgment call and we should look for "real-world" evidence which gives support to or contradicts the assumptions being made.

WHAT'S NOT IN THE ECONOMIC MODELS?

The economic models and the mathematics employed to obtain their valid conclusions can take us just so far, for much of what is going on is not reflected by the assumptions of these models. In a sense, less is to be found there than meets the eye.

An old Lithuanian saying (in free translation) asks: Where is the bone buried? The answer is to be found in interests, fears, experience, and a sense of priorities. Pro-NAFTA ads claim that 300 economists support NAFTA. A general point of view subscribed to by many economists running the spectrum from those considered conservative to liberal is that for the United States NAFTA is "economically trivial." No models suggest that the trade agreement will add much

more than $6 billion to America's gross domestic product over more than a decade, which is the amount that the American economy grows in a few weeks. So where is the bone buried?

INVESTMENT. Under NAFTA no investment from another North American country could be seized without full compensation. An investor would have the right to convert profits into another currency and take them out of the country. Investment in Mexico is viewed by many prospective investors as an opportunity to clean up in Mexico.

JOBS. Apart from the total number of jobs gained or lost, on which projections vary substantially, there is general agreement that the biggest impact of NAFTA will be on the nature of the jobs available. It is often said that "in the long run" we're better off. Those who struggle to get by in the "short run" are concerned; those whose jobs are not threatened in the short run can more easily afford to take a long term point of view. As the noted British economist John Maynard Keynes put it, "in the long run we are all dead."

THREAT OF JOB AND BENEFIT LOSS. Many in labor are concerned that the climate of NAFTA would strengthen management's hand in undermining worker wages and benefits through the threat of moving jobs to Mexico. Many in management like NAFTA for this reason. An old chess story comes to mind on this point. In a match between Alekhine and Vidmar, who favored big cigars, Alekhine summoned the referee to protest. But he's not actually smoking, replied the referee. He looks like he wants to, replied Alekhine, and as a chess player you know that the threat is more powerful than its realization.

CONSUMER SAFETY. Inspections of imported food and other products would continue, but there is concern that American safety standards could be eroded by challenges that they are unfair trade barriers.

ENVIRONMENTAL IMPACT. Environmental groups are deeply split on this with some arguing that the provisions of NAFTA adequately protect the environment and others arguing that we face an environmental disaster with NAFTA.

THE NATIONAL INTEREST. Many appeals have been launched on the basis of the emotionally appealing idea of "national interest." What is the national interest? Charles Wilson, who headed General Motors and held a high position in the Eisenhower administration, put it simply: "What's good for G.M. is good for America." The historian Charles A. Beard had more probing thoughts on the subject.[3]

WHAT DO THE NUMBERS TELL US?

"Mexicans bought 750,000 cars a year, but Detroit's Big Three sold only 1000 of them," fired Vice President Gore at Ross Perot in the opening minutes of their NAFTA debate held on 9 November 1993 and broadcast by CNN. The first salvo of numbers had been fired in the battle to convince the audience that he was authoritative because of the precision of his figures. In a salvo of his own, Mr. Perot charged that two-fifths of the exports to Mexico were phoney exports consisting of parts assembled into products returned to the United States. Mr. Gore countered with the claim that 80 or 90 percent of American exports physically stay in Mexico. The number ping-pong played in this debate epitomizes the kind of number slinging we have witnessed in discussions of NAFTA and domestic and international affairs in general.

Discounting the often felt desire to believe what we want to hear, which numbers are credible? In the case at hand, Mr. Gore's figure of 1000 cars sold by Detroit's Big Three to the Mexican market included only the number of cars shipped to Mexico from the United States in a year, not the number of cars supplied to the Mexican market by Big Three factories in Mexico. As to the percent of American exports that stay in Mexico, the figure was 68 percent in 1987 and 78 percent in 1992 according to the International Trade Commission, a bipartisan Federal agency.

The major lesson to be gleaned from these and earlier considerations is that numbers, so impressive because of the sense of precision they suggest, are slippery. It is untenable to view them as beacons of light standing as absolutes in a tumultuous world of conflicting interests and perceptions. They are tied to assumptions, conditions and qualifications which often appear in very small print or are lost altogether.

WHERE DOES THIS LEAVE US?

As to the pros and cons of NAFTA specifically and what is believable, this dear reader is for you to decide. Our objective here is to present an examination of the anatomy of an important case with many dimensions to help put into perspective what mathematical modeling and numbers can and cannot do for us.

Concerning where the bone is buried, the problem is that there are a number of bones buried all over, and to get a clear view of what is going on we must come to grips with them all. The tool provided by mathematical modeling to obtain an insight into the interplay of forces and how we can expect them to be resolved is effective to the extent that we can formulate "realistic" assumptions about the factors operating. These assumptions must be formulated in a way that allows us, using "appropriate" mathematical methods, to obtain their valid implications. These valid implications tell us what we can expect to occur when the underlying economic and social factors are played out. Whether the expectations derived will actually come to pass depends on the realism of the assumptions made. In the final analysis the devil is indeed in the assumptions.

REALITY'S VERDICT

As noted, before cutting a deal all concerned would be well advised to closely examine the assumptions which serve as its underpinnings. If a deal is cut, they would be well advised to follow up by examining its aftermath. By comparing what actually happens with what was predicted to happen under the assumptions made, we obtain reality's verdict on the credibility of the assumptions made.

The Hufbauer macroeconomic model (after Gary C. Hufbauer of the Institute for International Economics) predicts a gain of 170,000 jobs for the United States over the 1990s based on the fundamental assumption that Mexico runs trade deficits with the United States. The Prestowitz model (after Clyde V. Prestowitz, Jr. of the Economic Strategy Institute), on the other hand, predicts a loss of 420,000 jobs over the 1990s based on the fundamental assumption that Mexico runs trade surpluses with the United States.

NAFTA took effect on January 1, 1994, and in the first three months imports from Mexico to the United States grew much more rapidly than exports from the United States to Mexico. The United States still maintained a quarterly trade surplus with Mexico compared with the first quarter of 1993, but the gap closed by 45.1 percent to $560 million.[4] Both sides agreed, however, that the figures for one quarter were not a sufficient basis for determining a long term trend. But still, they give us some basis for reflections which, not surprisingly, differ considerably.

Mickey Kantor, the United States trade representative who had argued that NAFTA would create 200,000 jobs by 1995, declared victory by switching ground. He now saw the balance of trade as less important than its content and the increase in exports, which he viewed as key to raising America's standard of living. Some labor leaders saw the failure of U.S. exports to rise as fast as imports from Mexico as startling because Mexico had been protective of its markets whereas U.S. markets had been much more open. Whatever one's interpretation, the real outcomes are the barometers to be watched.

How free has free trade under the North American Free Trade Agreement turned out to be? As of June 1994 many exporters and importers on both sides of the border are disappointed; some feel that the agreement, in reality, is more appropriately described as a managed trade agreement or preferred trade agreement. United States officials have, however, argued that free trade does not mean open trade and that free trade cannot be achieved overnight.[5]

MATHEMATICS FOR ASTRONOMY

NEWTON SETS THE STAGE

In the late 1600s, a great argument took hold in London's scientific circles over what principle would account for the elliptical orbits of the planets. The astronomer Edmund Halley put the question to Isaac Newton who immediately replied, "the inverse square principle." That is, what is today called Newton's principle of universal gravitation: Every object in the universe attracts every other object with a force which varies directly as the product of their masses and inversely as the square of the distance between them. "How do you know?" asked Halley. "I calculated these orbits from this principle," replied Newton.

Newton presented these principles in his treatise, *Mathematical Principles of Natural Philosophy*[1] (1687). In his *Principia*, as it is referred to from its Latin title, Newton brilliantly synthesized the work of Copernicus, Kepler, Galileo and others, adding to it the product of his own genius, to formulate what we today term a mathematical model to describe the behavior of terrestrial and celestial objects in motion. He made fundamental advances in the mathematical instrument, called calculus, by means of which he obtained his valid deductions. What amazed Newton's contemporaries and leaves us in awe to this day is the far reaching scope of his work, its originality, and the close correspondence between the actual behavior of celestial objects and Newton's valid predictions. It took more than a century for others, distinguished in their own right, to refine and round out his work. Newton was modest about his accomplishment, noting that if he had seen further than others, it is by standing on the shoulders of Giants.

193

A NEW PLANET HAS BEEN DISCOVERED, BUT WHERE IS IT?

In the summer of 1801 the discovery of a new planet was excitedly reported by the press. This new planet, called Ceres, was first sighted on New Years Day by Guiseppi Piazzi at the Palermo Observatory. He was able to observe it only until February 11th, after which it was hidden by the Sun. Piazzi's attempts to compute the orbit of Ceres from the few position readings he had managed to obtain were unsuccessful.

Attempts were made by the many European astronomers to determine its orbit and predict the location of the new planet as it emerged from the Sun's cover, but without success. It seemed that Ceres had been lost.

AN ON-TARGET PREDICTION

Carl Friedrich Gauss (1777–1855), called the Prince of Mathematicians, was at the beginning of a long and distinguished career in mathematics and science at this time. Gauss, who had already achieved recognition for his work in number theory, took up the problem of determining the orbit of Ceres in November of 1801. He communicated his results to the astronomers Franz von Zach and Heinrich Olbers who located Ceres in the closing days of 1801 where Gauss' deductions predicted it would be found. This chain of developments added further support to Newton's theory of planetary motions, which provided the basic framework for Gauss' orbital calculation methods, and won Gauss recognition as a mathematical astronomer of the first rank.

Further observations of Ceres by Olbers led to his discovery of another planet, Pallas, in April of 1802 in the vicinity of Ceres. Gauss sustained his reputation as a mathematical astronomer by determining the orbit of Pallas. In 1804 a third planet, Juno, was discovered by Ludwig Harding and, as the recognized expert in such matters, the calculation of its orbit fell to Gauss as well. Ceres, Pallas, and Juno were the first discovered members of the asteroid belt, a numerous group of very small planets, called planetoids or asteroids, revolving about the Sun in the region between Mars and Jupiter.

MUDDLE OVER METHOD

In determining a planetary orbit from observations of its position, it is essential that these position readings be as accurate as possible. This is especially the case when the data are scanty, as was the case with Ceres. The difficulty is that if the same measurement is made several times, even under almost identical conditions, the results differ. The question that arises concerns how the values should be combined so as to obtain the result with the smallest amount of error in the "true" position of the planet.

The common practice in the case of a straightforward measurement, determining the length or diameter of an object, for example, is to take the average of the measurements obtained. In more complex cases in which two or more unknown quantities have to be determined, as when characteristics of the orbit of a planet have to be found from observations of the planet's position at different times, this method of taking the average is not applicable and a more general approach is needed.

Gauss developed a more general approach, called the method of least squares, while a teenager in 1795. This, he showed, reduces to the taking of the ordinary average in the case when a single unknown quantity (length or diameter of an object, for example) is obtained. He employed the least squares technique in his planetary orbit calculations and it shared the spotlight of success with his orbit determination methods. Gauss sent a summary of his results and methods to Olbers in August of 1802, but because of a curious sequence of developments they were not made widely known until 1809 with the publication of his great work on the *Theory of Motions of the Heav-*

enly Bodies[2]. A German version of *Theory of Motions* was completed by Gauss in 1806 and submitted to the publisher in 1807. But the Prussian army had been humbled by Napoleon in 1806, thus making the political situation very unstable. As a result the publisher agreed to accept Gauss' work only on the condition that he translate it into Latin, thus delaying its publication until 1809.

Independently of these developments, the French mathematician Adrien-Marie Legendre (1752–1833) formulated the method of least squares in an appendix to his book on methods for determining the orbits of comets[3], published in 1805. Gauss' remark in his *Theory of Motions* that he had used the method of least squares as early as 1795 greatly annoyed Legendre, who felt that Gauss was trying to take credit for his discovery. While this was not the case, public accusations were launched by an embittered Legendre which left the scene polluted. Olbers published the summary of Gauss' methods and results, including the method of least squares, which he had received in 1802 to set the record straight, but the damage had been done.

For the private Gauss, who shunned public controversy, it was an unpleasant episode which subsequently influenced him to keep out of the public limelight his views on non-Euclidean geometry. This was to have unfortunate consequences for the development of non-Euclidean geometry, which we consider in Chapters 17 and 18.

DÉJÀ VU

By 1840 six major planets of the solar system were known, Uranus, discovered in 1781, being the latest to be found. The orbit of Uranus was determined from widely separated observations of its positions and the expected deviations from its elliptical path due to the pull of the other planets were accurately predicted by Newton's principles. Uranus takes eighty-four years to complete a revolution of its orbit and for fifty years after having been discovered the planet behaved according to predictions. By 1830 significant differences between the actual behavior of Uranus and the path predicted for it by Newtonian mechanics had become evident. Reality and the mathematical model predicting its nature were in disagreement and with such confrontations the model always loses; it must back down. The question is, in what way?

There was the possibility, believed by some, that Newton's principles were not applicable over such large distances. Another possibility, which had its supporters, was that a hitherto undiscovered planet was exerting a gravitational pull on Uranus and causing the noted deviation from its predicted orbit. This possibility was independently taken up by John Couch Adams (1819–1892), while still an undergraduate at St. John's College of Cambridge University, and the French mathematical astronomer Urbain Leverrier (1811–1877). Adams solved the difficult mathematical problem of locating the position of this suspected planet from the observed motion of Uranus and Newtonian principles by September 1845. He determined the elements of the orbit of the suspected planet as well as its mass and position for 1 October 1845. Unfortunately, curious circumstances which fall under the framework of Parkinson's principle (whatever could go wrong will go wrong) led to misunderstanding and Adams was unsuccessful in having the Greenwich Observatory search the heavens for the suspected planet where his deductions predicted it to

be. If it had been searched for then, it would have been found very close to where Adams said it would be found.

Leverrier, who had an established reputation in the astronomical world, took up the problem and published his conclusions in a series of papers communicated between November 1845 and August 1846. The planet, subsequently named Neptune, was found by the Berlin Observatory on 23 September 1846, almost a year after Adams' prediction of its location. Adams was the first to obtain a mathematical deduction of Neptune's position, but Leverrier was the first to have his deduction verified by observation. Today the brilliant independent work of both mathematical astronomers is fully recognized. At the time a bitter dispute broke out over who had priority, illustrating again that the affairs of the planets, complex as they may be, are less tempestuous than those of mortals on Earth.

It was an extraordinary display of the power of mathematics—no data on the planet, but it must be there and there it is—which lent further support to Newton's mathematical model of planetary motion.

The orbit of Neptune was determined and the planet, which takes 164.8 years to complete a revolution in its orbit, was closely observed. In time the deviations of this planet and Uranus as well became larger than expected on the basis of known forces and the Newtonian model. An undiscovered planet was the cause of this behavior before and it seemed "reasonable" that a hitherto undiscovered planet was the cause in this situation as well. Percival Lowell (1855–1916) carried out the difficult deductions and after an arduous search the predicted planet, named Pluto, was found by the observatory in Arizona that he had founded.

Parkinson's principle was to have a role in this scene as well. The astronomer William H. Pickering had independently deduced Pluto's location in 1909 and had initiated a search for the planet at the Mount Wilson Observatory in California. It was not located, but after the discovery of Pluto by the Lowell Observatory in 1930 the old Mount Wilson photographs were reviewed and it was found that Pluto could have been found in 1919 had its image not fallen directly on a flaw in the photographic emulsion.

Thus another great triumph was scored by Newtonian mechanics, but trouble was brewing in the heavens for the Newtonian model in the behavior of Mercury and light.

TROUBLEMAKERS: MERCURY AND LIGHT

According to Newton's model for the planetary motions a planet revolves around the Sun in a stationary elliptical orbit, with other planets producing perturbations which could be calculated and observed.

It has long been known that the orbit of Mercury does not remain stationary as predicted by Newton's model. The path of Mercury is very nearly an ellipse, but it does not close up in one revolution. In the next revolution the path has advanced slightly in the same direction in which the planet was moving. The orbit is an ellipse which is itself slowly revolving as shown in Figure 14.1.

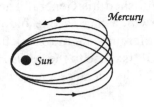

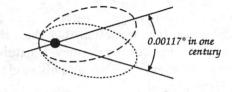

Figure 14.1   **Figure 14.2**

The amount of the shift is so small in terms of a year that it could not be detected, but if we let it add up for a century it comes to 0.00117 degrees per century as indicated by Figure 14.2.

How is the discrepancy between the observed shift in elliptical path and the prediction of Newton's model to be accounted for? Leverrier thought that the peculiar behavior of Mercury might be due to an interior planet which he tentatively called Vulcan. But, although thoroughly searched for, Vulcan was never found and Leverrier's hypothesis was never confirmed.

In 1916 Albert Einstein (1879–1955) published his General Theory of Relatively, a mathematical model which has as one of its theorems the deduction that the orbit of Mercury should shift by 0.00119 degrees per century. Einstein's result is close to the 0.00117 degree observed value, and his model entered the scene as a possible refinement of Newtonian mechanics.

Another clear point of disagreement between the Newtonian and Einsteinian models concerns the behavior of light. A ray of light bends when it passes near a "heavy" particle, such as the Sun. For a ray grazing the surface of the Sun, the numerical value of this deflection predicted by Einstein's model is twice that predicted by Newton's model. If the amount of curvature of a ray of light could be measured, we might emerge with a crucial result which tells us which model is closer to the mark. The successful determination of the measurement requires rather special conditions—a total eclipse of the Sun at a point where a large number of bright stars are in the background. Parkinson's principle held its fire and by good fortune such an eclipse took place on 19 May 1919. The results obtained supported Einstein's model.

These results gave strong support to Einstein's General Theory of Relativity as a refinement of Newtonian mechanics. Additional support in the form of experimental verification of the deductions from Einstein's model have been forthcoming. The Newtonian and Einsteinian models agree on many points. They disagree where "strong" gravitational forces are involved, for which Einstein's model provides the more accurate picture.

Another point of interest, further discussed in Chapters 17 and 18, is that Newton used Euclidean geometry as his picture of physical space whereas Einstein employed a non-Euclidean geometry.

MATHEMATICS AND CHANCE

THE ADVENTURES OF HASTY HARRY

What should I get my brother for his birthday, pondered bottom-line Bob. It's his thirtieth, the big 3, and this calls for something very special. His brother Hasty Harry, who was fond of games of chance, had an extensive collection of "unusual" dice and Bob decided to add to it by obtaining for him what would undoubtedly be the crown jewel of his collection, a gold die embedded with diamond chips to show off the spots on its faces. Bob had the die custom made and on the appointed day a very pleased Harry received a very special die.

Hasty Harry could hardly wait to show off his new treasure to his friends and, in addition, win some vacation money in a bit of friendly gaming activity that was certain to follow. In preparation for this he went to Martin's Models to obtain a probability model to describe the behavior of the new crown jewel of his collection.

"Martin, I want you to build me a probability model for the tossing of this die. I'm particularly interested in the probability that an even number shows."

"What can you tell me about your die, Harry? What do you know about its behavior from your experience with it?"

"I just got it as a present and I have no experience with it. I want to be prepared with a probability model before obtaining that experience. A die is a die, nothing special, apart from its being made of gold with diamond chips. What else is there to know?"

"All right Harry, I'll proceed on the assumption that it's an ordinary die of uniform composition, a fair die, as we say. This being the case, it is reasonable to expect all six faces to show up with roughly the same frequency, one-sixth of the time, when it's tossed a large number of times. This leads me to assign the same probability value of one-sixth to each of the six faces that can show. It follows that the probability or chance of an even number showing is one-half."

"What does this mean in terms of some friendly gaming activity, Martin?"

"If you toss this die a large number of times, an even number should show roughly half the time. We cannot say when an even number will show, that's a matter of chance, but it should show roughly 50% of the time."

REALITY STRIKES BACK

Harry proudly showed his die to his friends and all, with the exception of Harry, had a good time playing games of chance in which "friendly" bets were placed on which face would show when the die was tossed. Harry expected an even number to show roughly 50% of the time as predicted by Martin's model and bet accordingly. It came to pass, however, that after a large number of tosses of his die an even number had showed 67% of the time, which is sharply at variance with what Harry had expected. He had hoped to make a modest profit from this "friendly" gaming activity and now he found himself an unfriendly three hundred bucks in the red. Confused and feeling that he had been cheated, he stormed back to Martin's for some answers.

"I don't understand what went wrong Martin. If your mathematics is so precise, how could it happen that an even number showed 67% of the time instead of around the 50% you told me to expect? I'm three hundred bucks down because of this. You sold me a defective model and I want an explanation and my money back."

"Mathematics, the model I gave you in this case, Harry, delivered what it was capable of, namely, a valid conclusion with respect to the assumptions made. Please remember that it was you who provided me with the starting point of the analysis. I quote: 'a die is a die, nothing special.' As it turned out there was something very special about this die which made my assumption, based on your information, unrealistic. As a result we obtained a valid conclusion from the model about how often an even number can be expected to show which was at variance with the nature of your die. When there is a sharp conflict between a model and the reality it is to describe, reality always wins. Didn't your brother say anything to you about the nature of the die, or were you too dazzled by the gold and diamond chips to pay attention?"

"I'm not sure now. I'll have to ask him. Maybe I was too hasty."

A NEW MODEL FOR HARRY'S DIE

"Martin, I spoke to my brother and I listened this time. He said he told me that he had the die weighted internally so that the even numbered faces were twice as likely to show as the odd numbered ones. It's not at all an ordinary die in terms of its internal make up."

"I spoke to Bob after you left and he told me about the die's structure. He spent a lot of money to have the die made in this way and, ironically, it ended up costing you money. I made another probability model for your die based on the information Bob gave me. This model should be a much better fit to your die. It assigns the value 2/9 to the even numbered faces showing and 1/9 to the odd numbered faces showing, which fits Bob's information that the even numbered faces are twice as likely to show as the odd numbered ones."

"How much do I owe you?"

"The same as for the previous model. There's no additional charge for your hasty actions. You've paid that price already."

16 ▶ THE MATHEMATICS OF DATING

RADIOCARBON DATING

Earth is continually bombarded by sub-atomic particles, termed cosmic radiation, which are emitted by the Sun. These highly energetic particles react with atoms in the atmosphere to produce neutrons, which then collide with nitrogen atoms in the atmosphere to produce a radioactive form of carbon called carbon-14 or radiocarbon. Carbon-14 decays spontaneously, giving off an electron, and changing to nitrogen. This decay process is such that it takes about 5,730 years for half the amount of carbon-14 in a substance to disintegrate. This value is called the half-life of carbon-14. Carbon-14 behaves chemically in the same way as ordinary carbon-12.

Living plants and animals absorb carbon-14 along with carbon-12. When they die this absorption ceases, but the radioactive disintegration of carbon-14 continues, slowly at a fixed known rate. The simple idea that emerges, at least in theory, is based on measuring the proportion of carbon-14 left in a sample whose age is to be determined. Since the initial proportion of carbon-14 in the substance is known, to a good approximation, when it was living, we should be able to calculate how long the radioactive decay process had been going on, which gives us the age of the material.

Shortly after World War II Willard F. Libby proposed a way of employing the radioactive decay of carbon-14 to estimate the age of substances containing carbon-14, particularly organic remains. This includes charcoal, wood, cloth, limestone, bones, hair and soil. Libby's carbon-14 dating technique, for which he was awarded the

1960 Nobel Prize in chemistry, is based on the following assumptions:

1. Carbon-14 is produced in the atmosphere at an approximately constant rate.

2. Except for recent geologic time (the last half-century), there is a constant concentration of carbon-14 in all living things, which is about 1 carbon-14 atom to 1 trillion carbon-12 atoms.

3. The rate at which carbon-14 decreases is proportional to the amount present.

These assumptions lead to a mathematical model which gives us, as a valid consequence, a relationship describing the amount of carbon-14 in a substance in terms of time.[1]

Libby put his method to the test in 1955 when he obtained radiocarbon dates for a number of samples, mostly from Egypt, whose ages had been determined by other means. The radiocarbon dates were close to the established dates, and a powerful new dating tool was made available to archaeologists. This tool made possible decisive advances. Colin Renfrew notes that one of the greatest contributions of the first radiocarbon revolution was in making possible the study of world prehistory. Developments throughout the world may now be studied on a comparative basis with a sound framework of dates.[2]

Radiocarbon dating has undergone appropriate refinements. Libby's second assumption, for one, is open to question. The amount of carbon-14 in the atmosphere has not been constant with time. It has varied by as much as ± 5% because of changes in solar activity and Earth's magnetic field. In recent years contamination from the burning of fossil fuels and testing of nuclear weapons has resulted in significant changes in the amount of carbon-14 in the atmosphere. Studies of the bristlecone pine, a tree that grows in the White Mountains of California and lives for up to 5000 years, has allowed scientists to develop calibration curves for carbon-14 dates to correct for changes in the level of carbon-14 over time. Accurate tree ring records of age are available for a period as far back as 9000 years and scientists have sought other indicators of age against which carbon-14 dates can be compared. One such indicator is a uranium-thorium dating technique based on the decay of uranium-234 to thorium-230. Recently conducted studies at the Lamont-Doherty Geological Laboratory of Columbia University indicate that age estimates using carbon dating and uranium-thorium dating were in basic agreement for the period from 9000 years ago to the present. For earlier times the carbon-14 dates were substantially younger than those obtained by uranium-thorium analysis. The largest deviation, 3500 years, was obtained for samples that were approximately 20,000 years old. These results make clear some of the limitations of carbon dating and may lead to revisions in the age estimates that have been assigned by carbon dating to a number of "older" samples.[3]

A TALE OF FINE ART FORGERY

On May 29, 1945 the Dutch painter H.A. Van Meegeren was arrested on a charge of collaboration with the Nazis in having sold the priceless Vermeer painting "Woman Taken in Adultery" to Herman Goering. Van Meegeren claimed that he was not guilty of aiding the enemy in acquiring priceless Dutch art because he himself had painted "Woman Taken in Adultery," the famous "Disciples at Emmaus," as well as other paintings attributed to Vermeer and the less famous de Hooghs. The art world was understandably skeptical, and to prove his point Van Meegeren began, while in prison, to forge the Vermeer painting "Jesus Amongst the Doctors." When the work was nearly completed the charge of collaboration was changed to the less serious charge of forgery. Van Meegeren then refused to finish and age the painting in the hope of thwarting the investigation. An international panel of experts that was appointed to investigate the matter concluded that the alleged Vermeers were forgeries. Van Meegeren was vindicated and was sentenced to a year in prison for forgery. While in prison he died of a heart attack on December 30, 1947. In spite of the evidence gathered by the panel of experts, many people refused to believe that "Disciples at Emmaus" was a Vermeer forgery. It did not seem possible that such a masterpiece could have been forged by a painter considered third rate. More conclusive proof was demanded, and in 1967 scientists at Carnegie Mellon University took up the problem.

Their analysis was based on the radioactive decay of lead-210 and radium-226, small amounts of which are found in the widely used pigment lead oxide. Lead-210 has a comparatively short half-life of 22 years and if the paint used in painting is very old in comparison to this short half-life, the amount of radioactivity from the lead-210 in the paint will be approximately equal to the amount of radioactivity due to the radium; if the painting is recent, the amount of radioactivity from the lead-210 will be much greater than that generated by the radium. The Carnegie Mellon group made this precise by employing a mathematical model similar to, but more complex than, the one employed for carbon-14 decay. They showed that the amount of radioactivity from lead-210 is much greater than that from radium-226. In the end van Meegeren was in a sense vindicated; he was not a Nazi collaborator and he did paint a great work that had been attributed to Vermeer.[4]

THE MATHEMATICS OF SPACE

EUCLIDEAN GEOMETRY

Geometry, and more generally mathematics, as a system of postulates, definitions, and theorems or propositions deduced from the postulates, is the great contribution of the ancient Greek mathematicians collectively. To any such system they gave the name "Elements." Although not the first of its kind, the most successful Elements were those compiled by Euclid of Alexandria around 300 B.C.[1] Euclid set himself the task of taking the mathematics known at the time and organizing it into a deductive system. This involved deciding on which statements were to serve as the basic assumptions of the system (postulates), which were to be theorems, and providing proofs for the theorems, original ones when necessary. Euclid also had to decide how definitions were to be formulated. It was an enormous undertaking and, considering the decisive influence which his work has had, a brilliantly successful one.

Euclid took five geometric postulates as a basis for his deductive treatment of the geometry we now call Euclidean geometry and introduced them in the following manner.

Let the following be postulated:

I.   To draw a straight line from any point to any point.

II. To produce a finite straight line continuously in a straight line.

III. To describe a circle with any center and distance.

IV. That all right angles are equal to one another.

V. That, if a straight line falling on two straight lines makes the interior angles on the same side less than two right angles, the two straight lines, if produced indefinitely, meet on that side on which the angles are less than two right an-

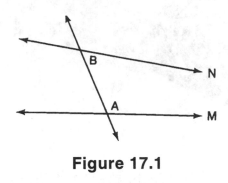

Figure 17.1

gles. That is, in terms of Figure 17.1, if the sum of angles A and B is less than 180°, then lines N and M will intersect at some point on the same side as angles A and B.

In more modern terminology, postulates I and II say: Through any two points, one straight line can be drawn; a line segment determines a line which is indefinite in extent. Postulate III says that a circle is determined when its center and radius are prescribed. Postulate IV says that all right angles are congruent. Postulates I-IV seem to express obvious truths in a simple fashion.

Postulate V, in contrast, is much more complex and as such stands in violation of the accepted criteria for postulates of the time, that they must express self-evident truths about spatial relations in a simple way. Euclid was aware of this, but part of his extraordinary accomplishment was to recognize that such a statement was needed to support the most complex of his geometric deductions and to boldly take it as a postulate after attempts to deduce it from simpler statements were unsuccessful. Euclid's fifth postulate is known as his parallel postulate, although the term parallel does not occur in it. He defines parallel lines as lines which being in the same plane and being produced indefinitely in both directions do not meet. The label parallel postulate is appropriate because it is equivalent to the following statement which involves the term parallel, equivalent in the sense that it plus the other four Euclidean postulates imply Euclid's fifth postulate, and vice versa. This equivalent to Euclid's parallel postulate is the one usually cited as Euclid's parallel postulate in textbook expositions of Euclidean geometry. It is known as Plairfair's postulate, after the Scottish physicist and mathematician

John Plairfair (1748–1819) who popularized it in a very successful textbook that he wrote on Euclid's Elements.

If given a line L and point P not on L, then there is one and only one line which passes through P and is parallel to L (Figure 17.2).

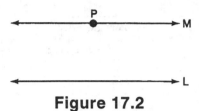

Figure 17.2

For the ancient Greek mathematicians and philosophers and their successors in mathematics, philosophy and science, Euclidean geometry served as a description of the space in which we live and as an intellectual discipline. It was not considered a mathematical model for space, with the admitted possibility that there might be other possible descriptions for physical space, but as the essence of physical space.

In more modern times the immense authority of the German philosopher Immanuel Kant (1724–1804), who viewed the postulates and theorems of Euclidean geometry as *synthetic a priori* propositions, strengthened this point of view. Propositions whose truth or falsity can be shown by reason, prior to observation, are called *a priori* propositions. A proposition such as "All apples are fruit" is an *a priori* proposition, but it gives us no factual information; its truth follows from the meaning of the words. Propositions whose truth or falsity follows from the meaning of the words are called analytic propositions. "All apples are fruit" is an analytic *a priori* proposition. A proposition which has factual content, such as "Some politicians pass slippery figures to the public," is called a synthetic proposition. Are there factual propositions whose truth can be established by pure reason, without recourse to observation? That is, are there synthetic *a priori* propositions? Kant argued[2] that the propositions of geometry and arithmetic are synthetic *a priori* propositions. They express factual statements, but they can be known by pure reason. The pro-Kantians believed that we have *a priori* knowledge of space; we do not know it from experience, but just the opposite; they are indispensable conditions to our having experience. Kant's views on the nature of mathematics gained wide acceptance in the late eighteenth and early nineteenth centuries, but this was to change radically by the last part of the nineteenth century.

A NON-EUCLIDEAN CHALLENGER

Euclid's contemporaries and successors greatly admired the organization of geometry which he had achieved in his *Elements*, but they were also dissatisfied with the price that had been paid in the form of a fifth postulate which could neither be considered simple nor self-evident. The parallel postulate problem that arose was to free Euclid from this blemish by either deducing the fifth postulate from Euclid's other postulates, or replacing it with an equivalent postulate which was simple and self-evident. The problem attracted many scholars from many lands. Some attempts to solve it were ingenious, but all were unsuccessful. In a departure from earlier approaches to the parallel postulate problem, which were direct, three mathematicians, working independently, brought a *reducio ad absurdum* or indirect approach to the problem. Each tried to show that Euclid's fifth was a consequence of his other postulates by showing that if the fifth or its equivalent is replaced by its negation, the amended set of postulates has implications that are contradictory. None were successful. The results they obtained were in contradiction to the nature of space as it was then perceived, but they were not in contradiction with each other.

In the early nineteenth century the parallel postulate problem was taken up by three men who reached startlingly different conclusions from their predecessors. Nicolai Ivanovich Lobachevsky (1792–1856), of the then recently estab-

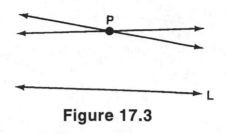

Figure 17.3

lished Kazan University in southern Russia, was first to publicly announce and publish his results. Lobachevsky took the contradiction of Plairfair's equivalent of Euclid's fifth in the following form: If given a line L and a point P not on L, there are at least two lines which pass through P and are parallel to L (see Figure 17.3). On the face of it this statement, which we shall term Lobachevsky's parallel postulate, seems absurd. In deducing the consequences of the amended system consisting of Euclid's postulates, with his fifth replaced by Lobachevsky's parallel postulate, Lobachevsky concluded that his system, now called non-Euclidean or Lobachevskian geometry, forms a consistent whole, although it was strikingly at variance with reality as it was then understood to be, especially in light of Kant's pronouncements on the nature of geometry and space. Lobachevsky first outlined his ideas in a paper which he presented at a meeting of the mathematics and physics division of Kazan University held on February 26, 1826. Three years later he elaborated on his ideas in his paper "On the Principles of Geometry," published in the Kazan Messenger. In the years 1835–1855 Lobachevsky further developed his non-Euclidean Geometry in a series of works.[3] He worked in virtual isolation.

He was appreciated by his colleagues as an outstanding teacher and administrator (having served as Rector of Kazan University from 1827–1846, and Assistant Guardian of the Kazan Educational District, (1846–1855)), but his ideas on geometry were incomprehensible to them and were treated with tolerance at best and derision and ridicule at worst. One of Lobachevsky's papers came to the attention of Carl Friedrich Gauss (1777–1855), who appreciated its worth and had Lobachevsky elected a member of the Gottingen Scientific Society in 1842. In letters to friends Gauss expressed the highest praise for Lobachevsky's work, but he never gave it public support. By the early 1820s Gauss had satisfied himself that a system based on a

denial of Euclid's fifth together with Euclid's other postulates is consistent but, apart from sharing his ideas in letters with trusted friends, kept his views to himself. Gauss was a private person who shunned controversy and had no desire to subject himself and his ideas to the wrath of the pro-Kantians. As he put it in a letter to one of his friends, he feared the "clamor of the Boeotians."

The basic idea of a non-Euclidean geometry arose from still another source. A Hungarian friend of Gauss, Farkas Bolyai (1775–1856), had spent much time trying to prove the parallel postulate from Euclid's other postulates. The problem attracted his son Janos Bolyai (1802–1860), who concluded that the system based on a denial of Euclid's fifth in the form, through a point not on a given line there are infinitely many lines passing through the point parallel to the given line, was free of contradiction. This system Bolyai called *The Science of Absolute Space.*[4] It was published in 1832 as an appendix to his father's text on geometry. Gauss' reaction to this work of the younger Bolyai was similar to the case of Lobachevsky, private approval but no public support. *The Science of Absolute Space* was Bolyai's one and only work on non-Euclidean geometry. Gauss' lack of public support and his discovery that Lobachevsky had anticipated him in publication left him so embittered that he did not further extend his ideas.

Some of the differences between Euclidean and Lobachevskian geometry due to the change in the parallel postulate are summarized in the following comparison table.

	Euclidean	**Lobachevskian**	
Given line L and point P not on L, there exist	one and only one line	infinitely many lines	through P parallel to L
Parallel lines	are equidistant	are never equidistant	
If a line inter- sects one of two parallel lines, it	must	may or may not	intersect the other
If two lines inter- sect in one point, there is	no line	one line	parallel to both
Two lines parallel to the same line	must	may or may not	be parallel
A line which meets one of two parallels	must	may or may not	meet the other
If two lines are parallel there are	infinitely many lines	at most one line	perpendi- cular to both
The number of circles passing through three non-colinear points is	exactly one	at most one	
The angle sum of a triangle is	equal to	less than	180°
The area of a triangle is	independent of	proportional to 180° minus	its angle sum
Rectangles	exist	do not exist	

Lobachevsky's far reaching investigations into what we now call Lobachevskian geometry convinced him that this system is mathematically legitimate in the sense of being consistent, that is, free of contradictory statements. While convincing, Lobachevsky's analysis fell short of being conclusive on this point. In 1868 Eugenio Beltrami (1835–1900) showed that Lobachevskian geometry is as consistent as Euclidean geometry by showing that the postulates of Lobachevskian geometry have an interpretation within the framework of Euclidean geometry where their conditions are satisfied. Such a structure is also called a model. Beltrami developed a model for Lobachevskian geometry within the Euclidean framework and in doing so showed that if an inconsistency existed in Lobachevskian geometry, then an inconsistency must also exist in Euclidean geometry. A mathematical Samson seeking to topple Lobachevskian geometry on the basis of inconsistency would also, if successful, topple Euclidean geometry; Lobachevskian geometry is thus as mathematically legitimate as Euclidean geometry.

Showing that a less familiar, controversial structure is as structurally legitimate as a more familiar one brings the less familiar one closer to us. But it also raises another question. Is the more familiar structure, or perhaps we should say, seemingly more familiar structure, as consistent as the less familiar one? Models of Euclidean

geometry have been given within the framework of Lobachevskian geometry, so that in the final analysis we can say that each is as mathematically legitimate as the other.[5]

WHICH ONE IS RIGHT FOR SPACE?

Since each is as consistent as the other, Euclidean and Lobachevskian geometries can peacefully coexist as mathematical structures very nicely. But can they coexist as rival models of physical space? If, like the pro-Kantians, you believe in physical space as being synonymous with a unique set of principles, Euclidean principles, then the answer is no; from such a point of view there is only one way to do it, the Euclidean way, and that's that. The relative consistency of Lobachevskian geometry in terms of its Euclidean counterpart makes it arbitrary, and thus indefensible, to choose one as a unique model for space on the basis of intellectual comfort. An examination of the postulates or theorems of each geometry against the behavior of physical space in terms of experimentation and observation emerges as the only satisfactory means for settling the issue of which geometry is the more realistic description of physical space.

The program that emerges entails assigning physical meanings to point and line and, in terms of these physical representations, empirically testing the realism of some key theorems of both geometries. An important property of Lobachevskian geometry is that for "small regions" it differs little from Euclidean geometry. For "sufficiently large" regions the differences become more substantial, but how large "sufficiently large" must be in order for significant differences to be revealed by experimental results is not clear. One key result in both geometries concerns the interior angle sum of a triangle. In Euclidean geometry this is 180 degrees. A possible approach suggested by this is to interpret point as the position of a celestial body, line segment as a beam of light projected from one such body to another, and determine the interior angle sum of large celestial triangles determined by three such celestial points. This is what Lobachevsky did in taking as his celestial points the Earth, Sun and star Sirius. He concluded that in this triangle the interior angle sum cannot differ from 180 degrees by more than 0.00000372". The result is inconclusive because we cannot answer the question of whether

Figure 93

the distances involved in this celestial triangle are sufficiently large to reveal deviations from Euclidicity. Another complication, which Lobachevsky recognized, is that the geometry used to describe space goes hand in hand with properties of matter in space. It might be that a discrepancy observed in the interior angle sum of a triangle from 180 degrees could be explained by retaining the assumptions of Euclidean geometry but at the same time modifying some physical assumptions involving mechanics or optics. By the same token the absence of a discrepancy might be compatible with the assumptions of non-Euclidean geometry and suitable adjustments in our assumptions about the behavior of matter in space.[6]

As to the application of geometry to practical measurements, engineering, surveying, and the like, "in the small," as we say, Euclidean geometry, suitably interpreted, is confirmed to a high degree of approximation as a theory of physical space. Lobachevskian geometry "in the small" approximates Euclidean geometry. Many of its principles, though not all, may be considered physically confirmed to a high degree of approximation. Engineers, surveyors and scientists who use Euclidean geometry in practice could use Lobachevskian geometry instead, but it is simpler to use Euclidean geometry since its formulas tend to be simpler.

18 ▶ A CHANGE IN THE CONCEPTION OF THE UNIVERSE

THE IMPACT OF THE NON-EUCLIDEAN CHALLENGER

Yet it was coming; and through that criticism of first principles which Aristotle and Ptolemy and Galen underwent waited longer in Euclid's case than in theirs, it came for him at last. What Vasalius was to Galen, what Copernicus was to Ptolemy, that was Lobachevsky to Euclid. . . . Each of them has brought about a revolution in scientific ideas so great that it can only be compared with that wrought by the other. And the reason of the transcendent importance of these two changes is that they are changes in the conception of the Cosmos.[1]

W.K. Clifford

The distinguished English mathematician William Kingdon Clifford (1845–1879) expressed this judgment in 1872 when the impact of non-Euclidean geometry was beginning to be felt. In an update Eric Temple Bell noted:[2]

The full impact of the Lobachevskian method of challenging axioms is probably yet to be felt. It is no exaggeration to call Lobachevsky the Copernicus of Geometry, for geometry is only part of the vaster domain which he renovated; it might even be just to designate him as a Copernicus of all thought.

Exaggeration or essence? This is not a simple question to come to grips with, but to appreciate what prompted these appraisals we

223

should look at the affect that the creation of non-Euclidean geometry
has had on how subsequent scholars view mathematics and science.

The geometric system perfected by Euclid came to occupy a
position of absolute authority for more than two thousand years. The
rules Euclid laid down for geometry of space were seen as inviolate
as the multiplication table. Space and Euclid had become synonymous. The creation of non-Euclidean geometry forced a profound
change in this point of view. The idea emerged that other geometries,
called non-Euclidean geometries, built from postulates differing
from those taken by Euclid, especially his parallel postulate, were
not only logically possible, but might be more realistic models of
space. The fundamental ideas that an applied geometry is a model for
space, not a structure inherent in space, and that experimental means
are the appropriate tools for deciding the issue of realism of its
propositions are a direct consequence of the development of non-
Euclidean geometry. Modern physics based on a non-Euclidean geometry only became possible as a consequence of this breakthrough.
The general mathematical modeling point of view explored in this
book is a development which emerged from this breakthrough.

The realization that postulates may be regarded as stated assumptions for purposes of further deduction, rather than supposedly self-evident truths, was slow in developing, but once it was established it proved to be a tremendous liberating force, a force unleashed by the successful launching of a non-Euclidean geometry. Mathematicians became free to study the implications of any assumptions which struck their fancy, no matter how fantastic and seemingly unrelated to the physical world they appeared to be, as long as they were internally consistent. Another outcome was that a new, more demanding level of deductive proof was brought to geometry in particular and mathematics in general, one which required that deductions stand on their own, independent of diagrams whose intent might be to indicate what is there, but which might also mislead us as to what is there.

A PROFILE IN COURAGE

It is uplifting to meet those who have the courage of their convictions and stand up for what they believe in, risking position, reputation and sometimes limb and life itself in the course of doing so. Lobachevsky was such a person. To challenge the accepted truths concerning Euclidean geometry in the early eighteenth century seemed as absurd then as proclaiming the flatness of the Earth would seem today. For a person with positions in science and education to do so would certainly, in the eyes of many, call into question his fitness to hold such positions. Such was Lobachevsky's reputation as teacher and administrator that his non-Euclidean "fantasy" was tolerated, albeit with ridicule and abuse. Looking back at the independence of mind he showed in his student years, the independence of mind of his mature years is not surprising.

Nicolai Lobachevsky was born on December 1, 1792 in Nizhi-Novgorod, Russia. His father, a minor government official, died when he was a small child and the family, already in difficult circumstances, was thrown into extreme poverty. Nevertheless, Lobachevsky's mother succeeded in moving her family to Kazan and managed to have her three sons enrolled as free scholars at the Kazan Gymnasium. Kazan University opened in 1805, and the university board made two decisions which were to profoundly affect Lo-

bachevsky's future. First, the board offered a university education at public expense to students completing the Gymnasium, under the condition that they devote six years to educational service after graduation. Lobachevsky's mother gladly accepted this proposal on behalf of her three sons, and Lobachevsky entered Kazan University in February of 1807. Second, the board invited teachers from Western Europe to join the faculty. Among those accepting this offer was Johann Bartels, a highly educated mathematician and gifted teacher. Lobachevsky took up mathematics under his guidance.

The Russian authorities considered universities centers of anti-government sentiments. Students were carefully observed. By a royal decree of May 8, 1811, students guilty of major offenses were to be expelled from the university and drafted into the army. Official reports on Lobachevsky characterize him as stubborn, repentless, ambitious, and manifesting signs of atheism. The latter was a particularly serious charge and only the vigorous protests of Bartels and other distinguished teachers saved Lobachevsky from having the royal decree applied against him.

In 1814 he was appointed assistant professor of mathematics and physics and in 1827 he was elected rector of the University. He was elected to this post six times and held it for close to twenty years, from 1827 to 1846.

Lobachevsky's views on education were unusual for the time and environment in which he found himself. In a speech given in July of 1825 at a general meeting of the University he declared:[3]

> You, whose existence chance has made a heavy taxation upon others, you, whose mind has been dulled and feelings deadened, you take no delight in life. For you nature is dead, the beauty of poetry alien, for you architecture has lost its charm and splendor, and the history of the ages is unentertaining. I take comfort in the hope that no such products of vegetable life shall leave our University, will not even enter it, if, unfortunately, they have been born with such a designation.

Lobachevsky died on February 24, 1856, almost thirty years to the day after the presentation of his initial paper on non-Euclidean geometry. His ideas, incomprehensible to the public during his lifetime, seemed headed toward oblivion.

Carl Friedrich Gauss died the previous year and his former students and colleagues undertook to publish his collected works. It was in this way that Gauss' views on non-Euclidean geometry became known to the scientific world at large. Gauss, publicly silent on non-Euclidean geometry during his lifetime, spoke authoritatively from the grave where he was beyond the reach of those he had termed the "Boeotians."

THE MATHEMATICAL MODELING PROCESS AS A TOOL FOR INQUIRY

OVERVIEW

As the situations considered in the previous chapters illustrate, the mathematical modeling process plays a fundamental role in helping us understand and live within the natural and institutional worlds in which we find ourselves. The following steps underlie the development and refinement of mathematical models for natural and institutional phenomena in all their diversity.

1. Specify as sharply as possible the task to be undertaken.

2. Collect relevant data about the situation in question and formulate an idealized representation for it. This is called a mathematical model (or more properly, the assumptions of a mathematical model), from which valid conclusions, called predictions, can be obtained by mathematical methods.

The setting of the phenomenon under study contains numerous features and characteristics, many of which are irrelevant or unessential to the focus of the study. In developing a model assumptions must be made as to which features of the setting are essential to the study and which are negligible. This calls for insight and judgment and is the most controversial aspect of the model development process; one person's essential might be another's irrelevancy.

Moreover, a delicate balance between realism and mathematical manageability must be struck. If we adopt a play-it-safe approach which favors incorporating into the model as many features as possible in the belief that we would thereby be less likely to leave out something important, we run the risk of making it mathematically intractable in that we might not be able to apply existing mathematical methods to obtain useful theorems from the assumptions of the model. If we dismiss too many features or the wrong ones as unessential to our study, then the model diverges too greatly from the phenomenon it is to represent and becomes unsuitable in this regard.

3.   Apply mathematical methods, the nature of which will depend on the situation, to obtain valid conclusions (theorems) with respect to the assumptions of the model.

4.   Test the accuracy of the model. That is, compare the valid conclusions of the model with the results obtained from reality by means of observation which may be coupled with experimentation.

If the valid conclusions can be shown to be true within some desired degree of accuracy, then this increases our confidence in the mathematical model as a realistic description of the phenomenon.

However, a model developed may not be the only representation for the phenomenon. Other models are possible as well. We cannot entertain the view of uniqueness in model representation. Albert Einstein and Leopold Infeld aptly describe this aspect of the model development process in the following terms:[1]

> In our endeavor to understand reality we are somewhat like the man trying to understand the mechanism of a closed watch. He sees the face and the moving hands, even hears its ticking, but he has no way of opening the case. If he is ingenious he may form some picture of a mechanism which could be responsible for all the things he observes, but he may never be sure his picture is the only one which could explain his observations.

Suppose alternative models become available for a phenomenon. How do we choose between them? The decisive verdict is rendered by reality. The crucial question is, are the model's theorems in accord with reality? When some of the model's theorems or predictions, as they are called, are found to be in disagreement with the findings of experimentation and observation, this tells us that some of the model's assumptions are incomplete or unrealistic and that the model must be refined or abandoned. A mathematical model lives as a description of a phenomenon as long as it avoids head on collision with some contrary fact of observation or experiment.

A theorem found to be false (or true for that matter) does not lose its status as a theorem. It remains a valid consequence of the assumptions of the model, correct in the sense of validity, incorrect in the sense of reality. It is unfortunate that many writers on science and mathematics use the terms valid and true synonymously. (Sometimes this is done out of sloppiness, sometimes out of ignorance.)

The following characteristics are also important for mathematical models. A model should bring in only observable entities connected with the phenomenon it is to describe. Although a model has its origins in data and observations, it must go beyond them in leading to significant new predictions and results. The predictions of the model should provide us with significant new insights into the phenomenon.

5. Refine the model as evidence obtained from experimentation and observation make necessary.

Figure 19.1 summarizes the basic steps in the development of a mathematical model in diagrammatic form.

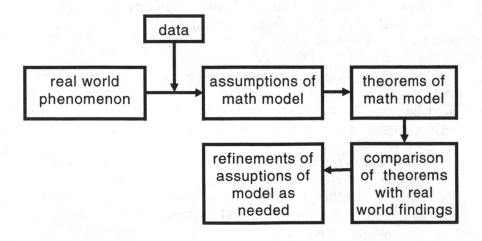

Figure 19.1

Some people find mathematics attractive because of what they perceive to be its absolutist nature. Mathematics is precise, they will say; there is no arguing with mathematics; you don't have ten sides to a story; mathematics provides us with unequivocal truth. Alas, mathematics does not provide a sanctuary for those seeking to escape from the difficulties of relativism. Anyone seeking a final expression of reality in a mathematical model is doomed to disappointment. No mathematical model can claim to express absolute and final truth. What we can hope to obtain is a more accurate description of reality through a sequence of refinements to a model. It is only when a model comes into conflict with some established fact of observation or experiment that the stage is set for the next refinement.

A classic illustration, discussed in Chapter 14, is provided by Newton's model of gravitation whose theorems confidently met the challenges hurled at them by reality for over two hundred years. One troublemaker was Mercury, whose orbit was found not to agree with the predictions of Newton's model. Attempts to reconcile the behavior of Mercury with Newtonian mechanics were unsuccessful. Newton's model of gravitation was succeeded by Einstein's model, his General Theory of Relativity, which accurately predicted the behavior of Mercury, the behavior of light, and accounted for the other behavior of nature as well as Newton's model. Einstein's achievement in refining Newton's model, we should observe, stands as one of the great landmarks of human thought. Nevertheless, when Einstein's model finds itself at odds with the behavior of nature it too will have to be refined.

The process of refining mathematical models for natural world and institutional phenomena is a central feature of the mathematical modeling process. R. E. Peierls offers the following insightful analogy to help illustrate the evolutionary nature of model building.[2]

> If we look at the photograph of a landscape in snow, showing part of a snowfield in sunlight and another part in the shade, we would describe this photograph as having one area in very light shade of grey, almost white, and another uniform area of a slightly darker shade of grey. On a more careful study of the same photograph under a microscope we discover that the grey areas are in fact made up of small black dots, the silver grains

of the photographic process, on a white background, the darker area differing from the lighter by having more of the black dots.

In a sense this discovery has proved the first description wrong, but it would be more reasonable to say that the new description refines the old one and replaces it when we are concerned with far greater detail than was the case at first. The old description is still good enough when we are concerned with taking a photograph or with looking at it. In fact by being simpler it is more valuable for that purpose. If we had to think of photographs always as collections of black dots on a white background we should find the photographs quite useless as pictures.

In the same way Newton's laws of mechanics and the wave theory of light retain their value for most practical problems within our everyday experience, in spite of our knowledge that the truth they contain is not the whole truth.

THE FINAL THEORY?

The idea of a "final theory," or equivalently "final mathematical model," expressed by some scientists who have achieved renown in their particular fields, is an appealing if untenable one.[3] The term final in such reflections is equated to comprehensive. Comprehensive theories, or models, of broad scope which bring together and provide unity to more and more seemingly diverse phenomena are meaningful to contemplate and desirable in the sense of enhancing our understanding of things. But the idea that such a theory could in any sense be final contradicts the inherent nature of a theory, or equivalently, mathematical model.

Simply and decisively put about his own theory and, by extension, all others, Albert Einstein had what is clearly the final word on the subject: "No amount of experimentation can ever prove me right; a single experiment may at any time prove me wrong."

20 ▶ GENERAL LESSONS AND OBSERVATIONS

FIT, FIT AND FIT

The accepted wisdom of the real estate business has it that the three most important considerations in buying real estate are location, location and location. In purchasing shoes the three top priorities might be said to be fit, fit and fit. In developing a mathematical model and employing mathematical methods for the study of a situation, be

it planning a route for a vacation trip or describing the structure of space, the three top priorities are the assumptions, assumptions and assumptions. Choosing realistic assumptions for the study of a phenomenon is as important as choosing realistic shoes to fit our feet. If the fit is poor in both instances, pain may be expected as a consequence. Little can be expected of mathematically derived conclusions as a suitable fit for the phenomenon under study, irrespective of the ingenious mathematical techniques and powerful computers employed, just as little can be expected in terms of comfortable shoes, irrespective of the fine quality leather and elegant style, if the model, or shoes, are poor fits.

THE CONSUMER'S FOCUS

As consumers of mathematics, be we students of a discipline which employs mathematical methods, managers of business enterprises who have been presented with recommendations based on mathematical analysis, or reflective folks who simply want to cut through what seems to be massive mathematical voodoo and computer hype to bedrock considerations, we are all led to the assumptions underlying the mathematical, statistical or computer analysis presented. Where is the bone buried, as we put it in discussing free trade in Chapter 13? The biggest bone is buried right there, in the assumptions. The questions that we should always be prepared to ask of ourselves, or others who are trying to convince us to adopt a course of action, are: What are the assumptions? Are they realistic? If others are trying to convince us to adopt a course of action, the burden of showing that the assumptions made are realistic is theirs. All we have to do is ask the right questions and listen for convincing answers.

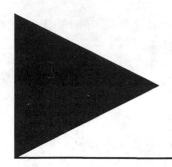

NOTES, REFERENCES, AND READINGS OF INTEREST

CHAPTER 1

[1]By multiplying 1 day = $86,400 by 11.57 and rounding off we obtain: 11.57 days = $1 million. By multiplying 1 year = $31.536 million by 31.71 and rounding off we obtain 31.71 years = $1000 million = $1 billion. By multiplying the preceding by 1000 and rounding off we obtain: 31,700 years = $1 trillion.

[2]For an English translation see D. Ferry, *Gilgamesh: A New Rendering in English Verse* (New York: Farrar, Straus & Giroux, 1992.)

CHAPTER 2

[1]F. Holzman, "Of Dollars and Rubles," *The New York Times*, Oct. 26, 1979.

[2]F. Holzman, "How C.I.A. Concocts Soviet Defense Numbers," *The New York Times*, Oct. 25, 1989.

[3]F. Holzman, "How C.I.A. Invented Soviet Military Monster," *The New York Times*, Oct. 3, 1991.

[4]F. Holzman, "C.I.A. Error Still Bloats Our Military Budget," *The New York Times*, Feb. 2, 1993.

[5]D. Stockman, *The Triumph of Politics: How the Reagan Revolution Failed* (New York: Harper & Row, 1986), p. 92.

[6]L. Barrett, *Gambling with History: Ronald Reagan in the White House* (New York: Doubleday & Co., 1983), p. 140.

[7]*Ibid.*

[8]Stockman, *op. cit.*, p. 98.

[9]R. Spector, Letter, *The New York Times*, Aug. 12, 1986.

[10]E. Richman, Letter, *The New York Times*, Aug. 12, 1986.

[11]J. Lehrer, *A Bus of My Own* (New York: Penguin Books, 1992), pp. 166–167.

[12]Many introductory statistics books contain a discussion of determining "real" dollar amounts that take into account inflation within a chapter on index numbers. See, for example, W.J. Adams, I. Kabus, M.P. Preiss, *Statistics: Basic Principles and Applications* (Dubuque: Kendall/Hunt Pub. Co., 1994), Sec. 14.5.

[13]Herodutus, *The History*; translated by D. Grene (Chicago: The University of Chicago Press, 1987), pp. 535–536.

[14]Population estimates of the regions of Greece are difficult to come by. A.W. Gomme, *The Population of Athens in the Fifth and Fourth Centuries B.C.* (1933), gives an estimate of 315,000 for Athens and the surrounding region in the age of Pericles (c. 450 B.C.). Estimates for the population of Sparta and the surrounding region vary from 182,000 to 376,000 c. 480 B.C.; see W. Durant, *The Life of Greece* (1939), pp. 74 and 682. E. Semple, *Geography of the Mediterranean Region* (1931), gives an estimate of 110,000 for Corinth, c. 480 B.C.

[15]"Smokings Annual Cost is Put at 5 Million Years of Life," *The New York Times*, Aug. 27, 1993.

CHAPTER 3

[1]J. Brooke, "In Brazil, Slip of the Tongue Makes Campaign Slip," *The New York Times*, Sept. 5, 1994.

[2]W. Styron, *Darkness Visible: A Memoir of Madness* (New York: Vintage, 1990). Also: "Prozac Days, Halcion Nights," *The Nation*, Jan. 4/11, 1993.

[3]G. Kolata, "Maker of Sleeping Pill Hid Data on Side Effects, Researchers Say," *The New York Times*, Jan. 20, 1992.

[4]T. Weiner, "Lies and Rigged 'Star Wars' Test Fooled the Kremlin and Congress," *The New York Times*, Aug. 18, 1993.

[4]T. Weiner, "General Details Altered 'Star Wars' Test'" *The New York Times*, Aug. 27, 1993.

[5]E. Schmitt, "Aspin Disputes Report of 'Star Wars' Rigging," *The New York Times*, Sept. 10, 1993.

[6]E. Chase, "The Tobacco Letters: Smoking, Health and Hypocrisy," *The Nation*, March 20, 1995, 382–387.

CHAPTER 5

[1]K. Robertson, "Oregon: Put Through the Mill," *The New York Times*, Nov. 27, 1989.

[2]F. Sutherland, "Why the Job Loss," *The New York Times*, Dec. 11, 1989.

[3]D. Stockman, *The Triumph of Politics: How the Reagan Revolution Failed* (New York: Harper & Row, 1986), pp. 123–124.

[4]To calculate the sum owed Senator Wind, call it S for the moment, we first note that S is given by:

$$S = 1+2+2^2+2^3+ \ldots +2^{20}$$

Multiplying through by 2 gives us:

$$2S = 2+2^2+2^3+ \ldots +2^{20}+2^{21}$$

Subtracting 2S from S gives us −S on the left side and $1-2^{21}$ on the right side since the 2, 2^2, 2^3, . . . 2^{20} values subtract out.

$$-S = 1 - 2^{21}$$

Multiplying both sides by −1 gives us:

$$S = 2^{21} - 1$$
$$= 2,097,152 - 1$$
$$= 2,097,151$$

CHAPTER 6

[1]D. Falkner, *The Last Yankee: The Turbulent Life of Billy Martin* (New York: Simon & Schuster, 1992).

[2]G. Will, "Paranoid in Pinstripes" *The New York Times Book Review*, April 5, 1992.

[3]J. Jagger, "Why Patriot Didn't Work as Advertised," *The New York Times*, June 9, 1991.

E. Marshall, "Patriot's Scuds Busting Record is Challenged," *Science*, May 3, 1991.

[4]M. Anderson, "The Reagan Boom—Greatest Ever," *The New York Times*, Jan. 17, 1990.

[5]W. Leontief, "We Can't Take More of this Reagan Boom," *The New York Times*, Feb. 4, 1990.

CHAPTER 7

[1]A. Bryant, "A Different Gauge for Rating Airlines," *The New York Times,* March 7, 1995.

[2]J. Davidson, W. Rees-Mogg, *The Great Reckoning*, rev. ed., (New York: Touchstone Books, 1994).

[3]R. Eisner, *The Misunderstood Economy* (Boston: Harvard Business School Press, 1994).

[4]R. Kuttner, *The End of Laissez Faire: American Economic Policy After the Cold War* (New York: Random House, 1991).

[5]C. McMillion, "Facing the Economy's Grim Reality," *The New York Times*, Feb. 23, 1992.

[6]NOVA, "Can You Believe TV Ratings?," WGBH, Boston, 1992.

[7]*Ibid.*

[8]C. Zwick, P. Lewis, "The Truth About the Deficit," *The New York Times*, Oct. 4, 1992.

CHAPTER 8

[1]M. Browne, "Coin-Tossing Computers Found to Show Subtle Bias," *The New York Times*, Jan. 12, 1993.

[2]"Hite/ABC Poll Comparison Analysis," news release by ABC for 6:30 p.m., EST, Oct. 26, 1987. Also, S. Squires, "Modern Couples Say They're Happy Together," *The Washington Post*, Oct. 27, 1987, Health Sec., p. 8, *"The Public Perspective,"* May/June 1990, p. 21.

[3]For further discussion of maximum likelihood estimates see, for example, W.J. Adams, *Finite Mathematics, Models, and Structure* (Dubuque: Kendall/Hunt, 1995), Ch. 7.

[4]For discussion of index numbers see, for example: W. J. Adams, I. Kabus, M. P. Preiss; *Statistics: Basic Principles and Applications* (Dubuque: Kendall/Hunt, 1994), Ch. 14.

[5]For an introductory discussion of the Consumer Price Index see P. J. McCarthy, "The Consumer Price Index," *Statistics: A Guide to the Unknown*, ed. J. M. Tanur et al. (San Francisco: Holden Day,

1972), 266–275; R. Samuelson, "How well does the Consumer Price Index measure the inflated prices we pay for dog food and doctors, parking lots and paperbacks?" *The New York Times Magazine*, Dec. 8, 1974.

CHAPTER 9

[1]R. Baker, "The Observer: Uncle Pete? Perot?," *The New York Times*, July 11, 1992.

CHAPTER 11

[1]For an introductory discussion of linear programming see, for example, W.J. Adams and R.B. Adams, *Get a Firmer Grip on Your Math* (Dubuque: Kendall/Hunt, 1996), Ch. 7; or W.J. Adams, *Finite Mathematics, Models, and Structure* (Dubuque: Kendall/Hunt, 1995), Ch. 3.

[2]L.V. Kantorovich, *Mathematical Methods of Organizing and Planning Production*, Leningrad University, 1939. For an English translation see *Management Science*, vol. 6, no. 4 (July 1960), pp. 363–422; or V.S. Nemchinov, ed., *The Use of Mathematics in Economics* (Cambridge, Mass: MIT Press, 1964).

[3]LP-1 is defined as follows:

Maximize $P(x,y) = 150x + 120y$ subject to:

$$x \geq 0, y \geq 0$$
$$2x + 3y \leq 1100$$
$$5x + 3y \leq 1400$$
$$4x + y \leq 756$$
$$x \geq 25, y \geq 40,$$

where x and y denote the number of DT-1 and DT-2 units, respectively, to be made per week.

[4]LP-2 is defined as follows:

Maximize $P(x,y) = 140x + 150y$ subject to:

$$x \geq 0, y \geq 0$$
$$8x + 5y \leq 2210$$
$$3x + 2y \leq 860$$
$$x \geq 50, y \geq 50,$$

where x and y denote the number of DT-1 and DT-2 units, respectively, to be made weekly.

[5]For details on this model see R.D. Buzzell, *Mathematical Models and Marketing Management* (Cambridge, Mass: Harvard Univ. Press, 1964), Ch. 5.

[6]P. Kotler, "Computerized Media Planning: Techniques, Needs and Prospects," *Occasional Papers in Advertising* (Urbana, Ill: American Academy of Advertising, 1965)

[7]F. Bass, R. Lonsdale, "An Exploration of Linear Programming in Media Selection," *Journal of Advertising Research*, vol. 3, no. 2 (May 1966), pp. 179–188.

[8]D. Gensch, "Different Approaches to Advertising Media Selection," *Operational Research Quarterly*, vol. 21, no. 2 (June 1970), pp. 193–219.

CHAPTER 12

[1]For further discussion see, for example, D. McNeill, P. Freiberger, *Fuzzy Logic* (New York: Simon & Schuster, 1992).

CHAPTER 13

[1]"For once, we're asking you not to believe in fairy tales," *The New York Times*, Nov. 13, 1993.

[2]"8 Fatal Flaws of NAFTA," *The New York Times*, Sept. 22, 1993.

[3]C. Beard, *The Idea of National Interest* (Chicago: Quadrangle Books, 1966; originally published in 1934).

[4]A. Myerson, "U.S.-Mexico Trade Advances Sharply Under New Accord," *The New York Times*, June 6, 1994, A1.

[5]A. Myerson, "Free Trade with Mexico? Not For All," *The New York Times*, June 21, 1994.

CHAPTER 14

[1]*Philosophiae Naturalis Principia Mathematica*; presented to the Royal Society in 1686 and printed in 1687.

[2]C.F. Gauss, *Theoria Motus Corporum Coelestium in Sectionisbus Conicis Arbientium* (1809). English translation: *Theory of the Motions of the Heavenly Bodies Moving About the Sun in Conic Sections* (1857); reprinted by Dover, New York, 1963.

[3]A-M Legendre, Nouvelles méthodes pour la détermination des orbites des comètes (Paris: 1805).

CHAPTER 16

[1]If we let y = N(t) denote the weight (in suitable units) of carbon-14 in a given portion of matter, where t is time measured in years, the mathematical model is expressed by the differential equation

$$\frac{dy}{dt} = ky,$$

where k is a constant.

By solving this equation and using the fact that the half-life of carbon-14 is 5730 years we obtain

$$y = N_0 e^{-0.0001209t}$$

as the function describing the amount of carbon-14 in a substance in terms of time, where N_0 is the amount of carbon-14 present at time t = 0 (the beginning of the process). Further discussion of technical details is found in many introductory calculus books; see, for example, W.J. Adams, *Fundamentals of Calculus with Applications* (Dubuque: Kendall/Hunt, 1996), Ch. 11. For discussion of Libby's work see W.F. Libby, *Radiocarbon Dating*, 2nd ed. (University of Chicago Press, 1955).

[2]C. Renfrew, *Before Civilization: The Radiocarbon Revolution and Prehistoric Europe* (New York: Knopf, 1973).

[3]M. Browne, "Errors are Feared in Carbon Dating," *The New York Times*, May 31, 1990.

[4]For further details see P. Coremans, *Van Meegeren's Faked Vermeers and De Hooghs* (Amsterdam: Meulenhoff, 1949).

CHAPTER 17

[1]For an English translation see T.L. Heath, *The Thirteen Books of Euclid's Elements*, 2nd ed., 3 vols (Cambridge Univ. Press, 1928; reprinted by Dover).

[2]I. Kant, *Critique of Pure Reason*, 1781, 1787.

[3]N.I. Lobachevsky, "Imaginary Geometry" (Scientific Papers of Kazan Univ., 1835); "Application of Imaginary Geometry to Certain Integrals" (Scientific Papers of Kazan Univ., 1836); "New Elements of Geometry with Complete Theory of Parallels" (Scientific Papers of Kazan Univ., 1835–38); *Geometrical Researches on the Theory of Parallels* (Berlin: Fincke, 1840). For an English translation see R. Bonola, *Non-Euclidean Geometry* (New York: Dover, 1955); *Pangeometry* (Scientific Papers of Kazan Univ., 1885); this work was dictated by Lobachevsky, then blind, in 1855, shortly before his death).

[4]For an English translation see R. Bonola, *Non-Euclidean Geometry* (Dover, 1955)

[5]See, for example, M.J. Greenberg, *Euclidean and Non-Euclidean Geometries: Development and History*, 2nd ed. (San Francisco: W.H. Freeman & Co., 1980); C.R. Wylie, Jr., *Foundations of Geometry* (New York: McGraw-Hill, 1964).

[6]For further discussion see N. Daniels, "Lobachevsky: Some Anticipations of Later Views on the Relation between Geometry and Physics," *Isis*, vol. 66, no. 231 (March 1975), pp. 75–85.

CHAPTER 18

[1]W.K. Clifford, "The Postulates of the Science of Space," Given before the British Association, 1872. Reprinted in J.R. Newman, *The World of Mathematics*, vol. 1 (New York: Simon & Schuster, 1956), pp. 552–567.

[2]E.T. Bell, *Men of Mathematics* (New York: Simon & Schuster, 1937), p. 306.

[3]V. Kagan, *N. Lobachevsky and his Contribution to Science* (Moscow: 1957), p. 62.

CHAPTER 19

[1]A. Einstein, L. Infeld, *The Evolution of Physics*, New Edition (New York: Simon & Schuster, 1961), p. 31.

[2]R.E. Peierls, *The Laws of Nature* (New York: Charles Scribner's Sons, 1956), p. 17.

[3]See, for example, S. Weinberg, *Dreams of a Final Theory* (New York: Pantheon Books, 1993); L. Lederman with D. Teresi, *The God Particle* (Boston: Houghton Mifflin Co., 1990); S. Weinberg, "The Answer to (Almost) Everything," *The New York Times*, March 8, 1993.

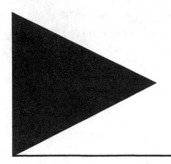

INDEX

A

absolute zero, 58, 59
acetaldehyde, 43
Adams, John C., 198, 199
Adams, John Q., 127
Adams, Ramuné B., 156, 244
Adams, William J., 156, 240, 243, 244, 247
advertising media selection problem, 173–175
AIDS, 67, 69
Alekhine, Alexander, 185
Alexander, Lamar, 77
Allen, Fred, 13
America Speaks, 114
American Federation of Teachers, 25
American Institute of Public Opinion, 114
American Medical Association (A.M.A.), 97
amoxicillin, 92
Anafranil, 53
Anderson, Martin, 88, 90, 91, 242
Andy's model, 163, 164
Angell, Marcia, 98, 99
Ann's model, 161–163
Arian heresy, 5
Aspin, Les, 51
assumptions, 19, 20, 22, 148, 161–165, 169–178, 182, 183, 189, 204, 206, 210, 236, 237
Austin Co., 168–170
Australia study, 67, 68
axiom, 162
AZT, 67, 69

B

Baker, Russell, 155, 244
Baltimore, David, 38
Barrett, Laurence, 19, 20, 240
Bartels, Johann, 226
Bass, Frank, 174, 245
Batten, Barton, Durstine, and Osborn (BBD & O), 173
Baumgartner, J.D., 66
Bell, Eric T., 223, 248
Beltrami, Eugenio, 220
Beard, Charles A., 186, 246
Bluestone, Charles, 92, 93
Boley, William, 96
Bolyai, Farkas, 218
Bolyai, Janos, 218
Bonaparte, Napoleon, 197
Bonola, R., 248
Bottom-line Bob, 170–172, 177–179, 203
breast implants, 41, 42, 62, 95–100
British parliamentary election, 120
Brooke, J., 240
Brown, Jerry, 86
Brown & Williamson Tobacco Corp., 44, 60
Browne, M., 243, 247
Bryant, A., 242
budget deficit, 111, 112
Bush, George, 29, 77, 86, 90, 106, 125, 126
Buzzell, R.D., 245

C

Campbell, S.K., 156
Cantekin, Erdem, 92, 93, 152
Carter, Jimmy, 19, 92
Cayonu, 5
Celsius scale, 58
Census of 1990, 135, 136
Centers for Disease Control and Prevention, 32
Centocor Co., 66, 67
Centoxin, 66–68
Central Intelligence Agency (C.I.A.), 14, 15, 152
Ceres, 194–196
Centrinet, 45
Chafee, John, 92
Chamorro, Violeta, 126
Chase, Edward T., 62, 241
Chatelperronian period, 5
Cheops, 5
Cherokee School, 49
Chow, Yung-Kang, 69
cigarette smoking, 32–33, 43–45, 60–63
Clay, Henry, 177
Clifford, William K., 223, 248
Clinton, Bill, 3, 6, 27, 29, 30, 86, 91, 125, 126
Colby, William, 15
Concorde study, 67, 68
Constantinople, 4
CONTAM report, 108
Coremans, P., 247
Council of Economic Advisors, 19
Cretaceous period, 6
Cro-Magnon man, 6
Crossley, Archibald, 115, 120

D

D'Amato, Alfonse, 22
Daniels, N., 248
Dateline NBC, 37, 154
Davidson, James, 104, 243
ddl, 69
Dean of Administrative Affairs, 12, 13
DeNoble, Victor, 43
DES (diethylstilbestrol), 69, 70

Dewey, Thomas E., 117, 118
Dewey versus Truman, 117, 120
Dinkins, David, 3
Dirksen, Everett M., 6
division by zero, 7–10
Dole, Robert, 29, 92
Dow Corning Corp., 41, 42, 96, 97, 100
Drolet, Robert, 86
Dukes, Graham, 40
Durant, W., 248

E

economic models, 182, 183
Egypt, 5
Einstein, Albert, 9, 10, 202, 230, 232, 233, 249
Eisner, Robert, 105, 243
Elias Sports Bureau, 84
Elwell, Ronald, 37, 152
Empire Blue Cross and Blue Shield, 47
epidemiological study, 97, 99
Euclid of Alexandria, 213
Euclidean geometry, 202, 213–215, 247
Euclidean versus Lobachevskian geometry, 219, 221, 222

F

Fahrenheit scale, 57, 58
Falkner, David, 84, 85, 242
Ferry, D., 239
Food and Drug Administration (F.D.A.), 39, 40, 45, 66, 68–72, 96–99, 153
Fourth Crusade, 4
Fox, Eugene, 51
Freiberger, P., 245
Friedman, Alexander, 9, 10

G

Gabriel, Sherine, 97, 98
Gallup, George, 114, 115, 118, 120
Garfinkel, Barry, 53
Gates, Robert, 17

Gauss, Carl F., 195–198, 217, 218, 227, 246
Geis, I., 157
General Accounting Office, 22, 52, 81, 82
General Motors case, 36–38, 62, 150, 152–154
Genghis Khan, 8
Gensch, Dennis, 175, 245
GIGO principle, 178
Gilgamesh, 5
Gingrich, Newt, 62, 76
Gomme, A.W., 240
Good & Evil, 108
Goodman, Edwin J., 55, 56
Goodman, Marvin, 17, 152
Gore, Albert, 186
Great Pyramid, 5
Greenberg, M.J., 248
Grundberg, Ilo, 39, 62
Gulf War, 85

H

Halcion, 38–40, 62, 150, 152
Halley, Edmund, 193
Handbook 8, 82
Harding, Ludwig, 196
Harper's index, 146
Harper's magazine, 146
Hasty Harry, 203–207
Heath, T.L., 247
Hebrew Exodus, 5
Herodotus, 31, 240
Hite, Shere, 132
Hite/ABC Poll Comparison Analysis, 133, 243
Hite reports, 132, 133
H.I.V., 69
Holzman, Franklyn D., 14, 17, 18, 239
Homer, 5
Hooke, R., 157
Hoover, Herbert, 116
Hufbauer model, 189
Huff, D., 157
Huxley College, 12, 102, 128, 137
        questionnaire, 57

I

Iliad, 5
Imanishi-Kari, Thereza, 38
In the Heat of the Night, 108
index number, 136–143
        consumer price (C.P.I.), 137, 142, 143
        I.Q. score, 137
        leading economic indicators, 137, 143
        price relatives, 140–142
        simple aggregative, 138, 141
Infeld, Leopold, 230, 249
interval scale, 58
interviewer induced bias, 126

J

Jackson, Andrew, 127
Jaffe, A.J., 157
Jagger, J., 242
Johnson & Johnson Co., 45, 46
Juno, 196

K

Kabus, Irwin, 240, 243
Kagan, V., 248
Kalidasa, 5
Kant, Immanuel, 215, 247
Kantor, Mickey, 190
Kantorovich, Leonid, 167, 168, 244
Kelsey, Frances, 70, 152
Kessler, David, 96
Kelvin scale, 58
Keynes, John M., 185
Kirk, James T., 185
Kolata, G., 241
Kotler, Philip, 173, 245
Kuttner, R., 243

L

Labor Department, 28, 110, 111
Landon, Alfred, 114, 115, 117
Lederman, L., 249

Legendre, A.-M., 197, 246
Lehrer, Jim, 26, 240
Leningrad Plywood Trust, 167
Leontief, Wassily, 90, 242
Leslie Fay Co.s, 46
Leverrier, Urbain, 198, 199, 201
Lewis, Peter, 111, 243
Libby, Willard F., 209, 210, 247
light, path of, 200, 201, 232
Liles, J., 157
linear program model, 169, 170
linear programming, 167, 173–175
Lobachevskian geometry, 217, 218
        as a model for space, 221, 222
        impact of, 223–235
        relative consistency of, 220
Lobachevsky, Nicolai I., 217, 225–
        227, 248
Lonsdale, Ronald, 174, 245
Lowell, Percival, 200
lupus, 97

M

Magna Carta, 4
Maldutis, Julius, 103
Marshall, E., 242
Martin, Billy, 84, 85
Martin's models, 203–207
math myths, 155
mathematical model, 162, 164–166, 189,
        203–205, 207, 221, 222, 231–234
mathematical modeling process, 229–
        234
mathematical precision, 170–172, 206,
        232
mathematics and the computer, 177, 178
Matlock, 108
maximum likelihood estimate, 134
McCarthy, P.J., 243
McMillion, Charles, 106, 107, 243
McNeill, D., 245
Mele, Paul, 43
Memorial Sloan-Kettering Cancer
        Center, 62
Mercury, 200, 201, 232
Mesopotamia, 5
method of least squares, 196, 197

Mexico, 181–186, 189, 190
Mondale, Walter, 124
Moseley, Shannon, 37, 62
Moses, 5
Mycenean culture, 5
Myerson, A., 246

N

Naim, John, 97
National Food Consumption Survey,
        81
National Institutes of Health, 38
National Medical Expenditures Sur-
        vey, 80
Neanderthal man, 6
Nemchinov, V.S., 244
Neptune, 199, 200
New Deal, 116
New York City Board of Education, 24
New York City Department of Investi-
        gation, 48
Newton, Isaac, 193, 246
Newton's principle of gravitation, 193
Newtons theory (model) of planetary
        motions, 195, 198, 199, 201, 202
Nicene Creed, 5
nicotine, 43
Nielson system, 108
Nixon, Richard, 92, 121
nominal scale, 58
Non-Euclidean geometry, 198, 202,
        217, 224, 225
nonresponse error, 116, 117
North American Free Trade Agree-
        ment, (NAFTA), 181–187, 190, 191
NOVA, 243
N.S.A.B.P. study, 52

O

Olbers, Heinrich, 195, 196
Oligocene period, 6
order of poll questions, 124, 125
ordinal scale, 58
Oregon, 74
Ortega, Daniel, 126

Oswald, Ian, 39
O'Toole, Margot, 38, 152

P

Paleolithic culture, 5
Pallas, 196
parallel postulate,
    Euclidean, 214, 215
    Lobachevskian, 217
    problem, 216
Patriot missile, 85, 86
Paulos, J.A., 157
Peierls, R.E., 232, 249
Pena, Federico, 37
Pentagon, 22, 51
Perot, Ross, 3, 105, 122, 186
Persian War, 32
Pétain, Henri, 156
Philip Morris Co., 43
photoceptor, 45
photopheresis, 45
pi (π), 55
Piazzi, Guiseppi, 194
Pickering, William H., 200
Pike, John, 51
Plairflair, John, 215
Pluto, 200
Poindexter, John, 62
Poisson, Roger, 52, 53
polls, 114–127, 155, 156
Poltrack, David, 108
postulate, 162
Preiss, Mitchell P., 240, 243
Prestowitz model, 189
probability model, 203, 204, 207

Q

quadrillion, 6
quota sampling, 118

R

radiocarbon dating, 209–211
random number generation, 131
random sampling, 118, 119, 128–130

Rasa's trip, 164–166
ratio scale, 58
Reagan, Ronald, 3, 29, 62, 75, 88–91,
    124
Reducio ad absurdum approach, 216
Rees-Mogg, W., 243
Reflective Ramuné, 170, 172, 177–179
Reich, Robert, 27
Reichard, R., 157
relativity theory, 9, 202, 232, 233
Renfrew, Colin, 210, 247
response options, 125
rheumatoid arthritis, 97
Richman, E., 240
Ricupero, Rubens, 36
Robertson, K.-A., 74, 241
Rome lab., 21–23
Roosevelt, Franklin D., 114–117
Roper, Elmo, 115, 121
Rudman, Warren, 104

S

Salamis, 31
sample size, 116
sampling methods, 118–120
Samuelson, R., 244
Schmitt, E., 241
Schulze, Charles, 92
Scud missile, 85, 86
Second Ecumenical Council, 5
selective service drawing, 129, 130
Semple, E., 240
Shakuntala, 5
silicone gel, 41, 42, 96, 97, 150
Slippery Statistics Society, (SSS), 147
Soviet defense outlays, 14–18, 151,
    152
Spector, R., 240
Spirer, H.F., 157
squaring the circle, 54
Squires, S., 133, 243
St. Augustine, 5
stability of voter opinion, 117
Stalin, Joseph, 168
Star Wars tests, 50–52
Stein, Herbert, 92

Stockman, David, 19, 20, 75, 151, 239–241
stratified sampling, 119
Styron, William, 39, 241
Sutherland, Frederic, 74, 241
Sylvester, Alice, 109

T

Talosian images, 35
Tanur, J.M., 243
target population versus the sampled population, 116
Temujin, 5
Teresi, D., 249
Thalidomide, 69–71
The Harrisburg Pennsylvanian, 127
The Literary Digest, 114–117, 119, 120
The Menagerie, 35
The New York Times, 60
theorem, 162, 164–166
Therakos, 45
Thermopylae, 31
Thurow, Lester, 105, 147
Time, 114
Trinitarian Christians, 5
Tripodi, Daniel, 45, 46, 152
Troy, 5
Truman, Harry S., 117, 118
Tsongas, Paul, 86
Turner, Stansfield, 14
Two-prime methylnicotine, 43

U

undecided vote, 118
unemployment rate, 110
Upjohn Co., 38–40, 62, 153
Uranus, 198, 200
U.S. News & World Report, 114

V

valid conclusion, 162, 164–16, 193, 206
Van Meegeren, H.A., 211, 212

Vermeer paintings, 211
vertex balding and heart attacks, 93, 94
Visigoths, 5
von Neumann, John, 131
von Zach, Franz, 195
Vulcan, 201

W

Waldo, C.A., 56
Weidenbaum, Murray, 19, 20
Weinberg, S., 249
Weiner, T., 241
Will, George, F., 84, 242
Williams, Merrell, 60, 61
Wilson, Charles, 186
Wine, Thomas, 60, 61
Wind, Lloyd L., 79
wording of poll questions, 121–124
Wylie, Jr., C.R., 248

X

Xerxes, 30, 31

Y

Yeaman, Addison, 44

Z

Zeno's paradox, 33
zero, 7–10
Zwick, Charles, 111, 243

Order Form

YES! I want _____ copies of *Get a Grip on Your Math*.
ISBN # 0-7872-1561-9

YES! I want _____ copies of *Get a Firmer Grip on Your Math*.
ISBN# 0-7872-1562-7

YES! _____I would like to inquire about other books written by
William J. Adams

Please call 1-800-228-0810 to order by telephone or Fax 1-800-772-9165. Prepayment is required.

☐ Check
    enclosed
☐ Charge my
    account

☐ Master Card ☐ American Express ☐ Visa

MC Bank #|__|__|__|__|__|__| Exp. Date ____/____/____

Account #|__|__|__|__|__|__|__|__|__|__|__|__|__|__|__|__|__|__|__|

Signature_____
(required for all charges)

Name_____

Phone (_____)_____

Address_____

_____

City/State/Zip_____

Please make check payable to:

Kendall/Hunt Publishing Company
4050 Westmark Drive
Dubuque, Iowa 52004-1840